The Open University

MU120
Open Mathematics

Unit 5

GW00644058

Seabirds

MU120 course units were produced by the following team:

Gaynor Arrowsmith (Course Manager)
Mike Crampin (Author)
Margaret Crowe (Course Manager)
Fergus Daly (Academic Editor)
Judith Daniels (Reader)
Chris Dillon (Author)
Judy Ekins (Chair and Author)
John Fauvel (Academic Editor)
Barrie Galpin (Author and Academic Editor)
Alan Graham (Author and Academic Editor)
Linda Hodgkinson (Author)
Gillian Iossif (Author)
Joyce Johnson (Reader)
Eric Love (Academic Editor)
Kevin McConway (Author)
David Pimm (Author and Academic Editor)
Karen Rex (Author)

Other contributions to the text were made by a number of Open University staff and students and others acting as consultants, developmental testers, critical readers and writers of draft material. The course team are extremely grateful for their time and effort.

The course units were put into production by the following:

Course Materials Production Unit (Faculty of Mathematics and Computing)

Martin Brazier (Graphic Designer)
Hannah Brunt (Graphic Designer)
Alison Cadle (TEXOpS Manager)
Jenny Chalmers (Publishing Editor)
Sue Dobson (Graphic Artist)
Roger Lowry (Publishing Editor)

Diane Mole (Graphic Designer)
Kate Richenburg (Publishing Editor)
John A.Taylor (Graphic Artist)
Howie Twiner (Graphic Artist)
Nazlin Vohra (Graphic Designer)
Steve Rycroft (Publishing Editor)

This publication forms part of an Open University course. Details of this and other Open University courses can be obtained from the Student Registration and Enquiry Service, The Open University, PO Box 197, Milton Keynes MK7 6BJ, United Kingdom: tel. +44 (0)845 300 6090, email general-enquiries@open.ac.uk

Alternatively, you may visit the Open University website at http://www.open.ac.uk where you can learn more about the wide range of courses and packs offered at all levels by The Open University.

To purchase a selection of Open University course materials visit http://www.ouw.co.uk, or contact Open University Worldwide, Walton Hall, Milton Keynes MK7 6AA, United Kingdom, for a brochure: tel. +44 (0)1908 858793, fax +44 (0)1908 858787, email ouw-customer-services@open.ac.uk

The Open University, Walton Hall, Milton Keynes, MK7 6AA.

First published 1996. Second edition 2008.

Edited, designed and typeset by The Open University, using the Open University TEX System.

Printed and bound in the United Kingdom by The Charlesworth Group, Wakefield.

ISBN 978 0 7492 2863 7

2.1

Contents

Study guide

This unit consists of five sections, the last of which is mainly devoted to helping you undertake your own statistical investigation, which comprises part of a tutor-marked assignment.

There is no material in the *Calculator Book* specifically linked to this unit. However, there is an optional chapter (Chapter 5), which provides a review of the statistical facilities which have been introduced in this block and which also gives an indication of some of the other statistical facilities available on the calculator.

Section 1 refers you back to *Units 2* to *4*, so it would be useful to have them handy. A short reader article is called on in Section 2. There is no audio band for this unit, but a video band (consisting of one long and one very short sequence) is used in both Sections 2 and 3. Both Sections 2 and 3 are quite long. If you do not have a DVD player available, you could study Section 4 before Sections 2 or 3. To complete one of the investigations in Section 3, you will need a tape-measure which is suitable for making measurements in a room.

You should now be becoming more experienced in scheduling study time for a unit. Think about the lessons you have learned in organizing and managing your study time so far. Can you state *one* thing you intend to do differently as you study this unit?

As you draw up a schedule for *Unit 5*, include your plans to complete the TMA question. Once again, do you intend to do anything differently from previous questions? How can you make improvements?

There are a number of Learning File activities in this unit. These are designed to help you stop and think about your progress so far, and plan for the next stage of the course. You will need to include time in your study schedule to complete them. At this stage, also think about your use of the Handbook work. As you come to new terms in this unit, make sure they are included.

On your planning sheet, as well as a 'timetable' for study and doing the relevant assignment questions, include the changes you intend to make and how you intend to go about making them. Think about these changes in relation to your own study situation, your progress on the course and to any particular areas you have identified for improvement. How will you know whether you have succeeded in achieving these changes? At the end of your study of this unit, come back to your plan and review your progress.

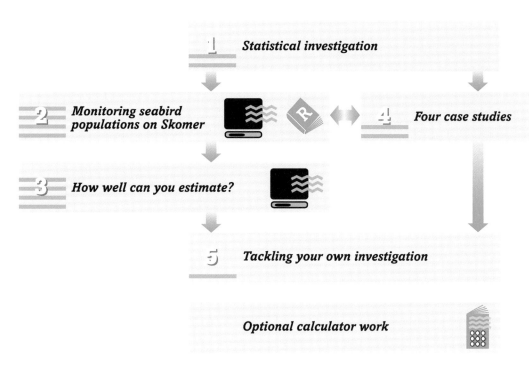

Summary of sections and other course components needed for *Unit 5*

Introduction

In this block, you have been introduced to a wide range of data-handling skills. However, it was stressed earlier that merely gathering data for its own sake is a relatively pointless exercise: you can simply end up drowning in a sea of data. In order to provide focus and direction for using these data-handling techniques, a four-stage model is proposed in this unit as a way for carrying out investigations.

The first four sections of the unit are directed towards giving you the opportunities to help prepare you for an investigation that you will carry out yourself in Section 5, which will represent a substantial part of your next tutor-marked assignment.

In Section 1, the stages involved in statistical investigations are discussed and you will be shown that most statistical investigations can be classified into one of three broad types. You will also have the opportunity to review some statistical techniques that have been introduced throughout the block.

In Section 2, the video takes you to the island of Skomer where you will get a chance to participate in a real investigation: monitoring the numbers in populations of seabirds on this nature reserve.

In Section 3, you will be given guidance in carrying out two simple investigations. The first involves you in assessing your own bird-counting skills. The second involves testing your estimation skills and investigating whether they improve with practice.

Four short case studies are provided in Section 4. They are drawn from a range of contexts and should give you a clearer sense of what is involved in carrying out and writing up a statistical investigation.

The final section of the unit asks you to tackle an investigation yourself as part of your tutor-marked assignment.

1 Statistical investigations

Aims   The main aims of this section are to introduce a four-stage model for carrying out a statistical investigation, to identify three broad types of investigation, and to review various mathematical and statistical techniques introduced in the block.   ◇

1.1 Decisions, decisions

How do you go about making decisions in your everyday world? Most people would find this a difficult question to answer—they just make them. Each decision seems very different and tends to be made on its own merits. Thus, it may not seem worth going to the trouble of applying formal methods to everyday choices and judgements.

Contrast this with how decisions tend to be made on behalf of large organizations—business, government and public services, for example. Here the scale and the consequences of the decision are much greater. The decision will affect not just the needs and desires of one person but of many hundreds or thousands of people. For reasons of fairness and public accountability, such large decisions should require greater thoroughness and formality in the process involved. Through discussion, both oral and written, the reasons for making a particular decision can be made explicit.

In a similar way, decisions and judgements in the academic world need to be convincing, not just to the individual making them but to the academic discipline as a whole. Within each discipline, processes have evolved for making judgements to ensure, as far as possible, that the decision process is both efficient and valid. *Unit 4* mentioned that with both 'data' and 'evidence', a 'quality' label can be attached.

It is the way of working towards making a decision which is of particular concern in this unit. In general, decisions are made on judgements which in turn may involve applying some criterion. However, despite apparent fundamental differences between the sorts of question, concept and vocabulary used in different disciplines, the ways of working are, fundamentally, similar. Whatever the context, and however the process is described, certain key stages are usually involved.

First, the purpose of the enquiry needs to be clear. This means that the central question must be clearly stated. Second, the investigator needs to think through what is involved and to collect appropriate information. The third stage involves analysing this information. The analytical tools required at this stage vary, depending on the nature of the enquiry and the level of sophistication of the investigator. Finally, any patterns in the results need to be interpreted making use of some plausible explanation of what is going on; why those patterns are emerging; what has caused them; what alternative explanations are possible; and so on.

These four stages are described below.

Stage 1   Pose the question

◁ *Posing* ▷   This is the crucial stage of any investigation, because it is the foundation on which all subsequent work is built. A well-posed question should be clear; that is, it should be worded simply and be as free as possible from ambiguity. Investigations where the central *question* has not been clearly formulated inevitably fall apart at a later stage. Note the stress on the word 'question' in the previous sentence. For any investigation to be worthwhile, it needs to have a clear purpose, and this is best expressed in the form of a well-thought-out question. For example, an investigation of the form 'Do traffic calming measures (such as 'sleeping policemen' or narrowing roads) work?' should not only yield useful conclusions, but also the subsequent statistical stages are likely to follow naturally.

On the other hand, a traffic survey for its own sake (simply counting the number of cars, lorries, other vehicles) will merely produce data with no enquiry to inform.

Too often, because the data are available, people embark on a statistical exercise by collecting large quantities of data without a clear sense of why, and end up drowning in a sea of data.

Stage 2   Collect the data

◁ *Collecting* ▷   With a clear focus on the central question, the next stage is to clarify what factors may be involved and to consider what data are needed, how they can be collected and any practical problems that this might entail—for example, equipment, location, measurement difficulties, accuracy issues, and so on. In fact, before rushing off to carry out a survey or set up an experiment, it makes sense to check whether the data you need have already been collected by someone else. Public libraries are stocked with excellent sources of data. Sometimes, however, the data you need are simply not available, in which case you must collect your own. Data you collect yourself are known as *primary* data. Examples of primary data might include a register of attendance, a list of prices or the results of a scientific experiment which you have carried out. Data which someone else has already collected are known as *secondary* data. Secondary data crop up in books, journals, magazines and newspapers: for example, government statistics, weather information, sports results, and so on.

Stage 3   Analyse the data

◁ *Analysing* ▷   There are many statistical diagrams and numerical computations available for analysing data and this is the stage where the calculator is invaluable. The techniques which comprise this stage are discussed in more detail in Subsection 1.2.

Stage 4   Interpret the results

Conclusions in statistical investigations are rarely watertight. There is usually scope for considering alternative explanations and raising new questions, as well as reviewing and evaluating what was done. Also, with the benefit of hindsight, you have the opportunity to evaluate the process that you went through and to consider whether it could have been done better. So, rather than this being the stopping point, you may find that this stage provides the stimulation to go through all or some of the stages again, this time with a new question or a refinement of the original question.

◁ *Interpreting* ▷

It may be helpful to think of these stages in a cycle as shown in Figure 1. This is because the interpretation stage is only meaningful if it attempts to answer the original question posed and because the interpretation stage very often leads to further questions with consequent loops around the cycle.

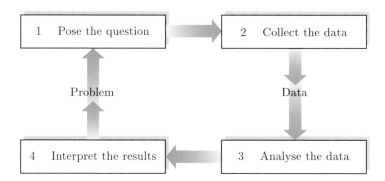

Figure 1   The main stages of a statistical investigation

The statistical investigation cycle is a simplified version of a more general mathematical modelling cycle that you will meet later in the course.

Once you have gained some experience of carrying out statistical investigations yourself, you may well find that thinking in terms of these four stages will help you to think through and plan your investigation in advance. For example, it is helpful to have thought through how you are going to analyse data before you start collecting them.

Recall Section 1 of *Unit 2* (the 'loaf of bread' example): it is not too difficult to pick out the key questions and issues in terms of these four stages, as follows.

Stage 1   Are people materially better off today than four-hundred years ago? In particular, how has the cost of a loaf of bread changed as a proportion of a daily wage?

◁ *Posing* ▷

Stage 2   You needed to know the price of a loaf of bread and typical wages for both 1594 and today. (This involved collecting both primary and secondary data.)

◁ *Collecting* ▷

Stage 3   The data were manipulated and arranged in a convenient form using percentages.

◁ *Analysing* ▷

◁ *Interpreting* ▷   Stage 4   Conclusion: the purchase of a loaf of bread represents a smaller proportion of a typical wage now than it did four hundred years ago. However, is looking at bread alone an appropriate way of answering the more general question about whether people are better off?

Activity 1   An example from Unit 3

Comments on Activities begin on page 65.

Recall Subsection 2.1 of *Unit 3* (the example concerning men's and women's earnings). See if you can identify each of the stages in the account. Write down a brief description of each of the four stages as they occur in the first part of this section.

Activity 2   An example from Unit 4

Recall Subsection 4.1 of *Unit 4*, 'A mother's tale'. Summarize each of the four stages as they occur in this example.

Can you think of another example of this cycle at work, either from the course or your own life?

1.2   Reviewing the techniques used at the analysis stage

This subsection looks in more detail at the analysis stage of a statistical investigation and in particular at the various statistical techniques that have been introduced in this block. By now you should be familiar with these techniques, but you might find it helpful to spend some time reflecting on when each is useful and what purpose it has.

Table 1 lists the techniques that have been described in *Units 2, 3* and *4*.

Table 1   Diagrams and computed values introduced in this block

Unit 2	Line graph (Section 1)
	Mean, weighted mean and median (Section 3)
	Ratios and indices (Sections 4 and 5)
	Pie charts (Section 7)
Unit 3	Range (Section 3)
	Interquartile range (Section 3)
	Boxplot (Section 4)
	Decile boxplot (Section 5)
Unit 4	Scatterplot (Section 1)
	Frequency diagram (Section 2)
	Standard deviation (Section 4)
	Relative spread (Section 4)
	Tally (Calculator Book, Chapter 4)

The next two activities ask you to review the progress you have made, both in understanding and in using these statistical techniques.

Activity 3 *Reviewing statistical techniques*

Allow at least thirty minutes for this activity.

Now consolidate your understanding of the terms and techniques in Table 1 by considering the contexts in which each of them was used earlier in this block. For example, the median was defined in *Unit 2* as the middle value of a batch of values, once they have been sorted in order of size, or the midpoint of the two middle values when the batch size is even. The median was used in *Unit 3* as a measure of average earnings by which different jobs and professions could be compared. One way of describing the essential purpose of a median, therefore, is that it is a single number which can act as a representative of the entire batch of data.

Table 2 lists these two properties of the median (its definition and its purpose). Likewise, the definition and purpose of an index have been included.

For each of the other techniques given in Table 1, try to write down, in your own words, a working *definition* and a statement of what you think might be the general *purpose* of using it. Look back at the definitions you have made in your handbook and use them to create a summary. You may find it helpful to write your summaries in a format like that in Table 2. Note that the summaries include only the key point(s) relevant to the term.

Table 2   A sample review of the techniques used in this block

Technique	Definition	Purpose
Median	The middle value of a batch of values, once they have been sorted in order of size, or the midpoint of the two middle values when the batch size is even.	A single number to act as a representative value of the entire batch.
Index	A number which measures how a value has changed over time, based on a starting value of 100.	A simple way of allowing comparisons to be made where things vary over time.

There is a printed response sheet for this activity.

Now that you have summarized and reviewed the mathematical techniques used in the block, it is a good time to think about your own learning of mathematics on the course. At this point think about the two questions set out below.

▶   Do you know which aspects of studying mathematics you are good at and which you are less good at?

▶   Have you a clear idea or strategy of how you are going to improve the latter?

The next activity asks you to assess your current skill level, to help you to identify evidence of your abilities, and to decide which areas you want to target for improvement.

Activity 4   A skills audit

First of all, think about the skills you have been using while studying so far. It is likely they will include many different areas such as using your calculator, using particular mathematical techniques, summarizing ideas, meeting deadlines, and so on. In identifying the skills you are using as you study MU120, you may wish to use the list you completed as you worked through the second audio sequence in *Unit 1*, other audit work you may have done, or the outcomes lists for each unit.

You might also find it useful to look at the categories of skill used in the N/SVQ core skills specifications in *Unit 1*.

To complete your own skills audit though, it is not sufficient just to identify the skills you are using; you also need to assess at what level you are using them. For this, it is useful to ask the question 'How do I know?' In other words, what examples and evidence can you use to back up your claim for using this particular skill at that specific level? Here again, the core skills material may be useful.

The final stage is to decide on your priorities for improvement. If you find there are a number of areas that you want to develop it may be sensible to prioritize them. For instance, you might plan to work on two areas as you complete Block B, and another one during Block C.

There is a printed response sheet which presents a format that you may like to use for your skills audit.

1.3   Three types of statistical investigation

The central theme of this unit is the carrying out of statistical investigations. One difficulty that people often have is that they understand how to perform and interpret most of the statistical techniques, but are somewhat at a loss when trying to decide which one to use in a real investigation and also when interpreting their results. You should find that your work on Activity 3 will pay dividends when you tackle such investigations. It should have given you a richer understanding of the general purpose and benefits of the various graphs and calculations.

There are three broad types of statistical investigation, each of which can be characterized by a simple question.

◇   How big is X?   (Summarizing)

◇   Is X bigger than Y?   (Comparing)

◇   Is X related to Y?   (Looking for relationships)

During the course of the next three sections of this unit, you will work through a number of such investigations. Each one can be classified into one of these three broad types according to the demands of the particular question posed.

You may be wondering why you should bother to classify investigations in this way. The main benefit of classifying your particular investigation is that you will be able to identify the techniques that are most likely to be of relevance when you get to the analysis stage. These three investigation types are now explained below.

Summarizing

Some investigations are simply of the form 'how many?' or 'how much?'. For example:

◇   How many people die from road accidents each day in the UK?
◇   How old are the students who enrol on MU120?
◇   At what speed do vehicles travel on a motorway?

You may have observed that these are not questions which have a single numerical answer. Each question requires a large number of measurements to be taken and each question will produce a batch of data as a result. Depending on the purpose of the question you are investigating, you may choose simply to express the results in a single summary value (perhaps using the median or the mean). Alternatively, you may also want to find some way of expressing the spread of values by calculating the interquartile range or the standard deviation. In most circumstances, a visual representation of the distribution of the data is useful, so some sort of frequency diagram or boxplot may be used.

Comparing

The most common reason for measuring things is to make comparisons. Here are some examples of 'comparing' investigations based on the three examples listed in the previous subsection.

◇   Do more people die from road accidents on weekdays or at weekends?
◇   Are students enrolled on MU120 older than students on an equivalent Arts course?
◇   Does motorway traffic in the UK travel faster than in France?

The key characteristic of this type of investigation is that two separate batches of data must be obtained. This may involve calculating the average of each batch and comparing them or, perhaps, drawing a boxplot for each set of values, placing one above the other, and comparing them.

Looking for relationships

Looking for relationships is where paired data are collected and you are interested in investigating whether there is a relationship between the two factors under consideration. For example:

◇   Are road deaths in different countries linked to their respective maximum speed limits?
◇   Is there a connection between the number of hours students work over the course of, say, a typical week and their final grade?
◇   Does life expectancy depend on the number of cigarettes smoked?

For this type of investigation, it is a good idea to plot the data on a scatterplot and look for a pattern. However, as was emphasized in *Unit 4*, the existence of a close association between two variables does not *prove* that one has caused the other.

Activity 5 *Classifying by investigation type*

To review the ideas of this section, classify the investigations below according to which type you think they are: summarizing (S), comparing (C) or looking for relationships (R).

(a) Do people with long legs tend to run faster than people with short legs?

(b) How much does a loaf of bread cost these days?

(c) Do men earn more than women?

(d) How heavy is a typical bag of crisps?

(e) Where two examiners have marked the same scripts, is there close agreement between the marks they award?

(f) Is there a link between poverty and ill-health?

(g) Do car seat-belts save lives?

(h) Are people taller than they were a hundred years ago?

Activity 6 *Review of the section*

This has been an important section in terms of introducing key ideas that you will build on during the rest of the unit. Spend a few minutes now summarizing the most important features of:

(a) the four *stages* of an investigation;

(b) the three *types* of investigation.

There is a printed response sheet which you will add to, as you work through the unit.

Outcomes

After studying this section, you should be able to:

◇   recognize the four-stage model of an investigation and identify
    and describe these stages in an account of an investigation
    (Activities 1 and 2);

◇   take an overview of the techniques used in the analysing stage of
    an investigation and have some insight into the general purpose of
    these techniques (Activity 3);

◇   recognize the three main types of investigation and classify and
    summarize investigations according to the categories (Activities 5
    and 6);

◇   discuss your progress on the course on the basis of a skills audit
    (Activity 4).

2 Monitoring seabird populations on Skomer

Aims   The main aim of this section is to illustrate the four-stage model in an actual investigation.   ◇

2.1 Four-and-twenty seabirds

In the previous section, the main stages in carrying out a statistical investigation were introduced, and emphasis was placed on the need to pose a well-defined question before embarking on the data collection and analysis parts of the investigation. In this section, you will be asked to look at a large-scale, long-term investigation: monitoring the population of seabirds on a Welsh island. For those carrying out the real investigation on the island, the four stages are not explicitly acknowledged, but nevertheless they provide an underlying structure for what the investigators do. In this section, the methodology is made explicit as each stage of the investigation is described. You will see how, after the interpretation stage, more detailed questions are formed, thus necessitating further loops around the cycle.

The investigation is set on the island of Skomer, which lies two miles off the south-west coast of Wales. This a National Nature Reserve, managed by the Dyfed Wildlife Trust.

Photo of Skomer cliffscape

Despite its small size, Skomer is internationally important for its seabirds and especially for its huge colonies of Manx shearwaters (160 000 pairs, 30% of the world population). The colonies of puffins, guillemots and razorbills are the largest in southern Britain and the lesser black-backed gull colony is one of the largest in Europe. The island has a resident warden, usually a number of researchers in addition, and there is accommodation for volunteer wardens. During the summer months large numbers of day visitors come to the island. The cliff scenery is spectacular, and during the spring and early summer months there are vast carpets of flowers, particularly red campion, bluebells and thrift.

During May and June, the main research activity on the island centres on counting the number of pairs of birds breeding on the island. Because there are many more of some species than others and because there are various types of nesting behaviour, different methods are used to count different species. This counting has been carried out with great care since the 1960s and the data collected allow the warden to monitor changes in population levels—an important factor in determining the management policy for the reserve. For example, does the ever increasing number of visitors cause problems for the birds during the breeding season? In what ways should the vegetation covering the top of the island be managed to provide suitable habitats?

Guillemots

In order to help you understand the particular problems involved in the counting process, you will be asked to watch a video sequence which was filmed during the breeding season. You will hear the warden, Steve Sutcliffe, talking about population monitoring and the methods used to count seabirds. After a general introduction, the video concentrates on a particular species of bird, the guillemot. These birds breed, packed closely together, on ledges on the cliffs.

While you are watching, be alert to areas where mathematical/statistical ideas can help. In particular, try to focus your thinking on what is involved in carrying out real statistical investigations and the four stages of posing the question, collecting the data, analysing the data and, lastly, interpreting the results. These words will appear in the top right-hand corner of the screen to remind you which stage of the investigation has been reached. After you have watched the video, you will be asked to carry out the analysis stage of an investigation into the number of guillemots before going on to interpret the results.

Watch band 4a of DVD00107 'Four-and-twenty seabirds'.

To provide further background to the guillemot investigation, there is a short reader article about guillemot populations which is taken from the *New Atlas of Breeding Birds in Britain and Ireland: 1988–1991* by Gibbons, Reid and Chapman (Poyser, 1993). The article discusses changes in the population of guillemots nationally and uses figures taken both from Skomer and other such colonies.

Now read the reader article 'Guillemot'.

Activity 7   More than the median

Notice the sentences in the second paragraph of the article:

> *Status of Seabirds* gives median colony size in England, Wales, and Ireland as 100 birds whereas in Scotland the figure is 400–500. Most of the birds, however, breed in much larger assemblies.

Make up a set of ten numbers representing the sizes of ten imaginary colonies in Wales for which the median is one hundred. Choose your figures to illustrate the fact that most birds breed in colonies larger than the median size.

Activity 8   Why do populations change?

The article suggests possible factors affecting guillemot numbers nationally. Note down some of these factors that may be affecting the Skomer population.

In the next two subsections, you will be asked to look in detail at the investigation described in the video sequence.

2.2   The total count

◁ *Posing* ▷   You will recall that the question posed in the video was in two parts.

How many guillemots are breeding on Skomer this year?

How does it compare with previous years?

On the video, you saw the data being collected. You may recall that, for the purposes of the investigation, the coastline of the island is broken up into forty-five sections. The sketch map of the island in Figure 2 shows the extent and location of each of these sections.

◁ *Collecting* ▷

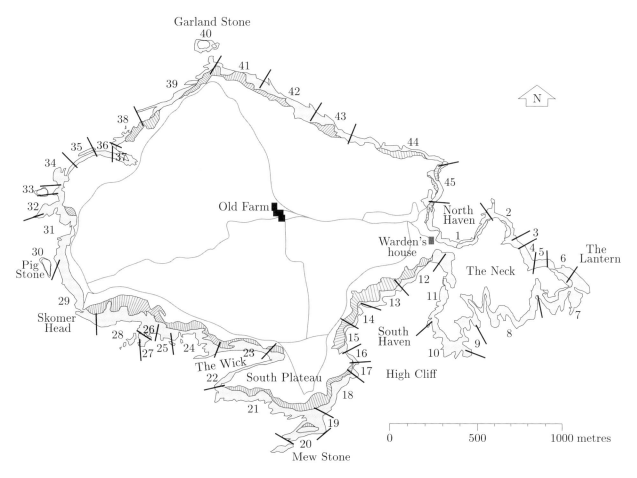

Figure 2   Skomer island sketch map with coastal sections marked

The data collected in 1993 are reproduced in Table 3 in their raw, unprocessed form. For each of the forty-five sections of cliff, every recorded count of guillemots has been listed. Notice that some sections of cliff have been counted more frequently than others. For example, ten separate counts have been made of the Pig Stone (section 30) and the numbers of birds counted on each occasion: 112, 108, 111, ... have been recorded.

The first three columns of the table give information about the forty-five sections of the cliff face, names (where they have them), and whether the birds can be counted from the sea (S) or from the land (L).

The fourth column gives all the recorded counts of guillemots. Some cliff sections had no guillemots, while not all sections were counted equally often.

The final column contains the medians of the data for most of the sections.

Table 3   Counts of guillemots on forty-five cliff sections in 1993

Section	Section name	S/L	Individual counts											Medians
1	North Haven	L	44	47	50									47
2	Rye Rocks	S	150	150	142	149	150	137	131	121	155	153		149.5
3	Protheroe's Dock	S	95	90	94	81	88	79	90	96	100	100	97	94
4		S	71	79	51	72	71	72	79	77	72			72
5	Shag Hole Bay	S	152	173	173	167	154	147	157	149	162			157
6	The Lantern	S	74	136	134	91	92	110	107	115	114	147	148	114
7	Little Sound	S	23	19	20	29	27							23
8	Matthews Wick	L	60	53										56.5
9	Castle Bay	L	58	70	50	52								55
10		S												
11	South Haven	L	15	30	37	38	35	32						33.5
12		L												
13		L												
14	South Stream Cliff	L	292	297	295	268	230	258	301	221	235	299		280
15		L												
16	High Cliff	L	708	728	746	744	697	723	722	689	736	785		725.5
17		S												
18		S												
19	Mew Stone W	L	15	15										15
20	Mew Stone S	S												
21		S												
22	The Wick	L	2431	2419										2425
23		L												
24	Wick Basin	L	60	63	38	38	50							50
25		L												
26	The Basin	L	39	39	44	40	40	36						39.5
27	The Amos	L	874	913										893.5
28	Tom's House	L	3	3	3	5	5	5	4					4
29		S												
30	Pig Stone	S	112	108	111	112	106	108	112	115	117	119		112
31	Pigstone Bay	S	156	159	147	151	150	164	181	182				157.5
32		S	75	75	76	85	85	69	76	76				76
33	Little Will Beach	S	475	466	490	495	457	469	475	466				472
34	The Table	S	230	226	268	269								249
35		S	110	94	147									110
36		S	324	263	266	263								264.5
37	Bull Hole	L/S	1258	1296	1216	1232	1310	1216	1251	1198	1204	1347		
38		S	184	190	195									
39	Payne's Ledge	S	226	219	230	199	191							
40	Garland Stone	S	88	88										
41		S	25	21	19	19								
42		S	197	160	160									
43		S	29	30	29	29								
44	Waybench	S	58	65	67	64								
45	North Castle	L/S	10	8										

Activity 9   *How many guillemots?*

(a) The final column of Table 3 shows the median number of birds counted in sections 1 to 36. Complete this column by calculating the medians for the remaining sections. Produce an estimate for the 1993 total of breeding guillemots on the whole island by adding the medians of all the forty-five sections of cliff face.

◁ *Analysing* ▷

(b) Remember that as you worked through the calculations you were dealing with the real, unprocessed data that had been recorded by the warden. You may feel uneasy with some of the data or you may feel that you would like to know more about them. If this is the case, make a written note.

(c) How accurate do you feel your answer is? Can you identify any inaccuracies that may lead to imprecise answers or errors?

The second question posed in this investigation was to determine how the 1993 bird population compares with previous years. The populations of guillemots and razorbills recorded on Skomer since 1963 are given in Table 4.

Table 4   Populations of seabirds on Skomer Island, 1963–1993

Year	Guillemots	Razorbills	Year	Guillemots	Razorbills
1963	7284	2996	1979	4923	2901
1964	7026	3377	1980	6965	3831
1965	7517	3459	1981	6789	3525
1966	7364	3089	1982	7067	3393
1967	6915	2550	1983	5112	3248
1968	6900		1984	6064	3335
1969	5888		1985	6181	3578
1970	3548	1893	1986	5835	3069
1971	4500	2421	1987	6192	2938
1972	5378	2310	1988	6532	2907
1973	4527	2094	1989	5556	2731
1974	5442	2468	1990	6051	2626
1975	5753	2448	1991	7516	2989
1976	5508	2232	1992	8032	3135
1977	5007	2336	1993	8696	3670
1978	5187	2358			

Notice that in some years the data are not recorded: maybe the data were not available. Or perhaps they were collected and then lost, or even jettisoned for some reason. The gaps are called 'missing observations'.

Activity 10   *Bird population trends*

(a) Use the data in Table 4 to draw a scatterplot on your calculator showing the variation in the guillemot population over the years 1963 to 1993. The years should be plotted on the horizontal axis and the population figures on the vertical axis.

◁ *Analysing* ▷

◁ *Interpreting* ▷ (b) Write a sentence or two to describe the way in which the guillemot population has varied over the years.

(c) Using the background information in the reader article 'Guillemot', and the statements made by the warden in the video, suggest some factors that may have affected the size of the guillemot population over the period.

(d) Produce a scatterplot on your calculator showing razorbill populations over the same period of time. Razorbills are the other species which breed in close proximity to guillemots and generally associate with them. Does your plot for razorbills show similar variations in population levels to the guillemot plot? How does this affect your confidence in the factors you identified in part (c)?

A razorbill

You now have answers to the two questions posed: you have an estimate for the population of guillemots on Skomer in 1993 and you are aware of how this compares with the population in previous years.

The analysis of the data in Tables 3 and 4 and the interpretation of the results raises several questions which have formed the basis of further investigations on Skomer. Two of these issues are considered in the next two subsections.

2.3   How do the counts vary?

◁ *Interpreting* ▷ The first issue that arises from the analysis carried out so far is the whole question of the mutual reliability or consistency of the counts or estimates. The number of birds on a particular cliff face is always changing. There is some variation during each day, as birds often stop incubating their eggs for a while. This occurs more at particular times of the day, and so counting is not done at those times. There is also some variation over the breeding season; sometimes an egg will fall from the cliff face and the parental pair will then leave; eggs and birds will suffer from predation by other species such as ravens. These factors mean that the population being counted is itself subject to variation.

Another factor contributing to the variation in the counts or estimates is a person's ability to count accurately. Inexperienced counters make mistakes such as counting birds twice, leaving out sections of cliff or mis-identifying birds. The ability of even the most experienced counters to count consistently depends on a number of factors such as the light conditions, how good a view can be had of the cliff face, and simple things like whether or not a particular bird is standing immediately behind its neighbour.

So, in order to get an idea of the accuracy of the estimates, a measurement is required of how much variation the estimates show. One explanation of this variation is the relative experience of the counters. However, even the best counters produce variation in their counts and it would be helpful to understand what causes this variation. One way of gaining an insight into some of the causes of variations in counts is to compare the consistency of guillemot counts made by one of the 'best' counters with similar measurements of consistency for another species such as the razorbill.

So the questions for this investigation can be posed precisely as follows.   ◁ *Posing* ▷

How much variation do the best counts of guillemots show?

How does this variation compare with that of the best counts of razorbills?

Jim Poole recording data at a cliff face

◁ *Collecting* ▷   In the video, you saw Jim Poole carrying out an accurate study of the guillemot colony at High Cliff. This is done in order to study the way in which particular ledges are colonized as the population changes—for instance, do the ledges become more densely packed or does the physical extent of the colony expand? However, these data may also be used to answer the questions outlined above.

Jim's counts are taken to be the best that can be achieved. That is, it can be assumed that variation due to measurement error has been reduced to a minimum. He is a most experienced counter; he uses a sketch map of the cliff face (shown in Figure 3), counting and recording each subsection separately; he uses a telescope from exactly the same position on ten successive days in June (weather permitting).

Jim's recordings of his data for numbers of guillemots on each subsection (A–L) of High Cliff during 1993 are given in Figure 4, together with his calculation of summary statistics and his data and statistics for razorbills for the whole cliff, carried out at the same time.

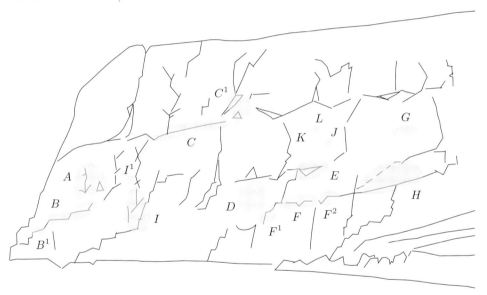

Figure 3   Sketch map of High Cliff

Activity 11   *How consistent?*

◁ *Analysing* ▷   (a) Use your calculator to find the mean and standard deviation of the daily totals for guillemots (bottom row of the main table) and for razorbills given in Figure 4. Compare your answers with the statistics given in the figure: which divisor has been used in calculating the standard deviation?

Record the median count and interquartile range for each of the two species.

(b) What do these figures tell you about the relative variation in the recorded counts for the two species?

STUDY PLOT COUNTS 1993

GUILLEMOTS   HIGH CLIFF

	2/6	3/6	4/6	5/6	6/6	7/6	8/6	9/6	10/6	13/6
A	44	45	44	41	43	46	51	47	49	53
B	40	37	40	42	37	36	34	34	36	40
B'	16	15	14	12	14	16	13	11	14	15
C	109	106	113	110	109	109	112	104	103	113
C'	5	4	5	6	5	5	4	5	4	5
D	114	119	128	124	114	119	126	122	115	131
E	21	25	24	24	24	23	26	26	20	29
F	51	58	63	57	65	62	58	55	57	65
F'	20	29	29	32	23	24	23	24	22	26
F²	6	11	9	9	8	7	8	8	8	9
G	12	10	13	12	12	11	13	12	13	14
H	221	217	211	219	209	212	202	193	208	226
I	26	31	29	32	25	27	28	25	23	33
I'	4	5	5	5	4	6	5	5	4	6
J	8	5	7	6	6	6	8	7	7	8
K	8	8	9	9	10	9	8	8	9	10
L	4	3	3	4	3	3	2	3	3	4
	708	728	746	744	697	723	722	688	736	785

Range 689–785 $\bar{x}$ = 727·8

σ_{n-1} = 27·6

RAZORBILLS

2/6	3/6	4/6	5/6	6/6	7/6	8/6	9/6	10/6	13/6
172	166	179	154	156	164	167	153	177	188

Range 153–188

$\bar{x}$ = 167·6
σ_{n-1} = 11·6

Figure 4 Study plot counts, 1993 (Jim Poole's own data and summary statistics)

You have figures for the variation in counts of the two species, but there is a problem with comparing them directly. A variation of, say, 20 around an average of 167 is, in relative terms, far more important than a similar-sized variation around an average of 726. What is needed in a case like this, where the two medians themselves differ widely, is a statistic which measures *relative* rather than absolute variation. One such statistic, which was introduced in *Unit 4*, is the *relative spread*. This is found by dividing the interquartile range by the median and converting the result to a percentage.

Activity 12   *Relative spread*

(a) Using your results from Activity 11, find the relative spread for each species.

(b) What do your two answers tell you about the relative variation in the counts for guillemots and razorbills?

◁ *Interpreting* ▷   The relative spread of the razorbill counts is more than twice that of the guillemot counts. But how is this result to be interpreted?

Remember that Jim Poole made these counts at the same time and from the same position. Why then should there appear to be twice as much relative variation for one species as for the other? Do razorbills tend to come and go more frequently than guillemots? Unfortunately, it is not clear: even though the two data sets were obtained under identical conditions, as far as possible, measuring error has not been eliminated completely because of the different behaviour patterns of the two species. As Jim pointed out in the video, razorbills tend to take positions hidden away in the crevices of the rock, making them much more difficult to locate than the guillemots, which, although they bunch together, are more likely to be visible on the exposed rock ledges.

Another possible approach is to look for any patterns in the two sets of numbers to see if this provides clues to the source of the variation. One way of doing this is to consider the counts of guillemots and razorbills made on each of the ten days in June as paired data and to investigate whether there is any relationship between them.

Activity 13   *Looking for patterns*

◁ *Analysing* ▷   (a) Use the daily totals from Figure 4 to display on your calculator's screen a scatterplot of razorbill numbers compared with guillemot numbers. Plot guillemot numbers on the horizontal axis with razorbills on the vertical axis, and plot one point for each day's total.

(b) Describe, and try to interpret, any noticeable patterns in your plot.

◁ *Interpreting* ▷   Looking at the two sets of data alone revealed little about any possible relationship between the counts of the two species of birds on different days. It is only when a scatterplot is drawn that a clear sense of the relationship emerges. Looking at the scatterplot, you can see that the points lie in a broad band running from bottom left to top right. This suggests that, by and large, the greater the number of guillemots recorded on any particular day, the greater the number of razorbills also recorded. For example, on 13 June the highest numbers were recorded for both species, and on 9 June the lowest numbers occurred. The only obvious exception to this pattern was on 5 June when a low razorbill count coincided with a relatively high guillemot count.

It is likely then that there may be some other factor causing the numbers of both species to fluctuate together at High Cliff; perhaps temperature or wind speed. Alternatively, fluctuations in numbers may be related to disturbance to the breeding colonies caused by human visitors or natural predators. Yet another statistical investigation cycle of posing, collecting, analysing and interpreting might be employed at this point to try to identify whether any of these factors are affecting the number of birds sitting on eggs.

In fact, investigations of this type in the past have led to an understanding that wind speed is likely to be an important factor. If the wind speed is greater than force 4 on the Beaufort scale , birds tend to be unsettled on the breeding ledges, and their numbers constantly vary making the counts unreliable. As a result, no counting of guillemots or razorbills is carried out on Skomer in very windy conditions.

In this subsection, questions have been posed about the variation in guillemot and razorbill counts. Comparing relative spreads of counts conducted in identical circumstances suggests that razorbill counts are more variable than guillemot counts. However, certain conditions occurring on particular days may well affect the counts of both species.

Guillemots on a breeding ledge

2.4   Is the increase in population evenly distributed over the island?

In Subsection 2.2, the investigation concluded that there had been a large increase in the population of breeding guillemots on Skomer in 1993 and that this appeared to be part of an ongoing trend over recent years. A question which arises from this, which may have implications for the warden's management policy, concerns whether this increase is spread evenly across the various colonies of the island. For example, it could be the case that larger colonies are expanding at a faster or slower rate than the smaller ones.

◁ *Posing* ▷   So the questions can be posed precisely as follows.

> Is the growth in the guillemot population between 1992 and 1993 distributed evenly across the various subcolonies?

> Is there any link between colony size and rate of population growth?

◁ *Collecting* ▷   The data needed for this investigation have already been collected. Table 5, opposite, gives the median figures for the forty-five sections of cliff face: those which were collected in 1992, together with those for 1993 some of which you calculated in Activity 9. Both figures are given correct to the nearest whole number.

The first two columns of the table give information about the forty-five sections of cliff face including their names (where they have them). Columns 3 and 4 give the average (median) numbers of guillemots recorded in 1992 and 1993. In column 5, the increase in guillemot numbers is shown, with a negative number being used where the numbers decreased from 1992 to 1993. Column 6 shows that increase, expressed as a percentage of the 1992 figure. For example, the average number of birds recorded at Tom's House (section 28) fell from 8 to 4, a reduction of 4 which is a percentage increase of $^{-}50\%$.

◁ *Analysing* ▷   The increase in guillemot numbers and the percentage increase (that is, the rate of growth) for each section have already been given in the table, correct to the nearest whole number.

▶ What technique might you use to find out whether there was a relationship between the size of the subcolony and its rate of growth?

Table 5   Increase in guillemot numbers, 1992–3

Section	Section name	1992	1993	Increase	% increase
1	North Haven	48	47	⁻1	⁻2
2	Rye Rocks	134	150	16	12
3	Protheroe's Dock	97	94	⁻3	⁻3
4		56	72	16	29
5	Shag Hole Bay	177	157	⁻20	⁻11
6	The Lantern	113	114	1	1
7	Little Sound	19	23	4	21
8	Matthews Wick	66	57	⁻10	⁻14
9	Castle Bay	59	55	⁻4	⁻7
10					
11	South Haven	34	34	0	0
12					
13					
14	South Stream Cliff	187	280	93	50
15					
16	High Cliff	722	726	4	0
17					
18					
19	Mew Stone W	9	15	6	67
20	Mew Stone S				
21					
22	The Wick	2298	2425	127	6
23		2	0	⁻2	⁻100
24	Wick Basin	41	50	9	22
25					
26	The Basin	31	40	9	27
27	The Amos	701	894	193	27
28	Tom's House	8	4	⁻4	⁻50
29					
30	Pig Stone	116	112	⁻4	⁻3
31	Pigstone Bay	176	158	⁻19	⁻11
32		74	76	2	3
33	Little Will Bench	460	472	12	3
34	The Table	184	249	65	35
35		118	110	⁻8	⁻7
36		224	265	41	18
37	Bull Hole	1062	1242	180	17
38		178	190	12	7
39	Payne's Ledge	202	219	17	8
40	Garland Stone	107	88	⁻19	⁻18
41		26	20	⁻6	⁻23
42		204	160	⁻44	⁻22
43		33	29	⁻4	⁻12
44	Waybench	71	65	⁻7	⁻9
45	North Castle	5	9	4	80

We chose to use a scatterplot to help explore whether there is any such relationship. This is shown in Figure 5.

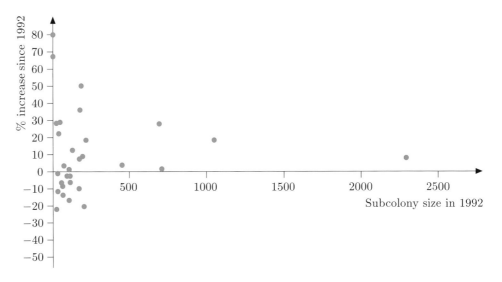

Figure 5   Scatterplot of percentage increase in size against size in 1992 for all subcolonies

▶   Can you see any pattern in the scatterplot?

◁ *Interpreting* ▷

◁ *Analysing* ▷

There does not appear to be any relationship between the size of a subcolony and its increase in size since the previous year. However, many of the points are crowded together, so it might be useful to look more closely at that part of the plot.

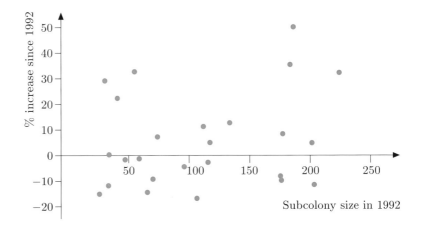

Figure 6   Scatterplot of percentage increase in size against size in 1992 for subcolonies with size between 25 and 250

◁ *Interpreting* ▷

Again there does not appear to be any relationship between percentage increase and colony size. There is certainly no evidence here that, for example, large subcolonies grow any faster or slower than small ones.

▶   So what did you think when a pattern did not 'jump out' at you?

After all the work involved in the analysis, it is perhaps disappointing to find that there does not appear to be any noticeable relationship. However, this may itself be a useful and interesting result, encouraging a search for other possible patterns of population growth. For example, there may be some relationship between the rate of growth of the various subcolonies and their physical position—the subcolonies on the cliff sections on the north coast all seem to have decreased in size. Yet another statistical investigation seems to suggest itself, with more posing, collecting, analysing and interpreting!

2.5   Matters arising

In this subsection, a number of issues which have arisen during the course of the investigations in this section will be discussed.

Measurement inaccuracy in a dynamic population

When carrying out some measurements, such as the width of a desk, the object being measured is unchanging and stays still, so any variation in measurements recorded must be entirely due to the method or instrument used to carry out the measurement—and, of course, perhaps due to you, the measurer. This variation linked to measurement inaccuracy is to be expected even though you may try to be as accurate as possible.

In measuring a population that is changing, there are two sources of variation: measurement inaccuracy is still present, but there is also the actual variation inherent in a changing population. Often, it is this very change in the population that you wish to measure, but there are then difficulties separating what you wish to measure from measurement inaccuracy. For example, if the warden wishes to determine whether daily fluctuations in the population are linked to the number of visitors, he must find a way of ensuring that the fluctuations he is measuring are not themselves due to measurement inaccuracy. Similar issues were apparent in Subsection 2.3 when the different variations in guillemot and razorbill counts were considered.

Problems caused by improving measuring techniques

Suppose it were to be found that the total counts could be made more accurate by, for example, only counting during particular times of day when the birds are most settled. This might result in counts which were subject to less measurement inaccuracy, and therefore better estimates.

However, since this improved technique was not available in previous years, care would have to be taken when comparing population levels with those calculated by a different technique earlier. It may be that what is considered most important is the year-on-year comparison, so the new, more accurate, techniques may have to be rejected. Alternatively, both techniques could be used for one year to ensure some continuity.

Rounding issues

In Activity 9, you were asked to comment on how accurate your total for the island was and to identify possible sources of error. Look back now at what you wrote at the time.

The sum of the medians was 8694. Corresponding calculations with the means produces 8730, but you probably would not have felt that either was exactly right. It might have been safer to say that the figure was approximately 8700, and certainly nearer 8700 than 8600 or 8800.

Why then are most of the figures for the previous twenty years given so accurately? Are you really expected to believe that the number of guillemots present in 1963 was 7284 and not 7285 or 7283?

c. is an abbreviation and stands for the Latin word *circa* meaning 'about'.

These population figures were drawn from summary sheets filed in the warden's office. In some cases, the counts for each of the sections of cliff are present. In other cases, it is just the final figure (for example, 7364) that appears. On the 1968 sheet, the figure '*c.* 6900' is shown. Does this mean that the warden at the time was unable to make detailed counts and just made a reasonable estimate? Or did he do the count in the usual way and then approximate the total to the nearest hundred? There is no obvious way to find out. On the other hand, with the 1966 figure, 7364, you can be pretty certain that the warden did as accurate a job as possible and *you* could decide to round it. The problem is that without experience of doing the actual count, it is difficult to decide on the appropriate level of accuracy to which to round. Should it be 7360, 7400 or 7000? Therefore, the way you present your results needs to reflect these decisions.

Another matter to bear in mind is what it feels like to discard accuracy when you have invested a considerable amount of time and effort into achieving as much precision as possible with the counting. It is easy to imagine successive wardens, after weeks of data collection and hours of analysis, eventually coming to a final figure and finding it very hard to say, for example, 'So my best estimate is 7364 but I had better approximate that to 7400'. However, if the result is to be made widely available, and is to appear credible, then either appropriate rounding must be carried out or the best estimate given, accompanied by an indication of the precision.

Messy data

When analysing data from real investigations, problems are often encountered. In Activity 9, you were also asked whether you felt uneasy with some of the data in Table 3 or whether you felt you would like to know more about them. You may well have felt uncomfortable with some of the data. For example, two of the largest cliff faces, the Wick and the Amos, appear to have been counted only twice. Was this really so? Why could this have been? You were not in a position to be able to do anything other than make do with the data you had and this is often the case in an actual investigation. In fact, the two figures that you were given were averages produced on two different days by several observers, working together. They divided the cliff face into sections, produced averages for each and added them together. This raises a more general question about data collected in previous years on the island and, indeed, about secondary data in general. Without access to the people who did the counts or a record of their activities, you have no way of knowing exactly how the measurements were taken, whether or not they were averaged, and so on. This is worth bearing in mind when *you* carry out an investigation.

Another example of messy data occurred in the list of guillemot counts for the cliff face called the Lantern (section 6). The list is:

74   136   134   91   92   110   107   115   114   147   148.

There is quite a wide spread of values here but to the warden the number 74 looked suspiciously small. This count is described as an 'outlier', as it appears to lie a long way out from the main body of the data. It could have been due to an error by the counter or perhaps this count was taken when large numbers of birds had left the nest ledge for some reason. When the warden came to analyse these data, he doubted the validity of the number 74 sufficiently to discard it completely. Again, this process, often called 'discarding outliers', is something which is often necessary in real investigations.

Plotting all the values on a scale like the one shown in Figure 7 is a good way of both encapsulating the overall variation in the counts and identifying outliers. Perhaps if the warden had used this technique he might have been less certain that the 74 was an outlier.

▶   Do you think the warden was justified in discarding the 74?

Figure 7   Guillemot counts on the Lantern cliff face

A common example of messy data is where real measurements and tables of data have missing values. An example of this occurred in Table 4 with the yearly razorbill estimates. There were no figures recorded in 1968 and 1969, presumably because no counts were carried out in those years. Sometimes it is possible to remedy missing data by further data collection, but often (as here) it is necessary to make do with what you have. In this section, you have looked at part of a real, large scale, statistical investigation, monitoring the population of seabirds on Skomer. The counting of one particular species, the guillemot, was chosen and you have seen how the methodology introduced in Section 1 underlies the work of monitoring the population of this species.

After attempting to answer the questions posed in Subsection 2.2, you went on to investigate the amount of variation in some of the counts. This was followed by an attempt to determine whether the increase in population was evenly distributed over the island.

The monitoring of seabird populations carried out on Skomer provides many examples of investigations based on the four-stage model. These dealt with here were chosen to illustrate the use of the statistical techniques and concepts that have been covered in the course so far, as well as providing an opportunity to discuss a number of important issues which frequently arise in the course of a real investigation.

Activity 14   *Investigating investigations*

Skim back through the section, paying particular attention to the margin icons (Posing, Collecting, Analysing, Interpreting), and how they relate to the various statistical investigations carried out on Skomer.

Look back at the notes you wrote for Activity 6 at the end of Section 1. In the light of what you have read in this section, do you need to add any comments about the most important features of the four stages of such an investigation?

In particular, do you wish to add any summary notes from the issues discussed in Subsection 2.5 'Matters arising'?

There is a printed response sheet for this activity.

Outcomes

After studying this section, you should be able to:

◇ understand that the monitoring of seabird populations on Skomer consists of a range of statistical investigations (Activities 8 and 9);

◇ describe how statistical investigations can be broken down into the four stages of posing the question, collecting the data, analysing the data, interpreting the results (Activity 14);

◇ describe to someone else some of the issues associated with 'real' statistical investigations (Activities 9, 11, 13 and 14);

◇ carry out an analysis of given data and interpret the results (Activities 7, 9, 10, 11, 12 and 13).

3 How well can you estimate?

Aims   The main aims of this section are to guide you through two statistical investigations, drawing attention to the four stages, and to revise and extend skills introduced in previous units of this block.   ◇

3.1 Estimation, approximation and types of data

The two short investigations outlined in this section are both concerned with estimation, an important mathematical skill which people generally seem to acquire by experience. For example, when people first begin to cook, they often follow recipes precisely, weighing out the ingredients accurately. Once they become more experienced, they acquire an ability to estimate half a pint of milk or 200 g of flour. Not only are they able to make the estimation, but they also become more aware of when estimation is appropriate and what level of accuracy is acceptable. They probably know, for example, that the quantity of oil used when making mayonnaise is crucial, whereas the amount used when frying an onion is not so important.

How far do you recall estimation being part of your previous formal learning of mathematics? Can you remember being encouraged to estimate the length of a line or the number of objects collected on a table, for instance? Certainly in traditional mathematics classes, estimation and approximation were not stressed a great deal. Until the 1980s and beyond, children were taught 'long multiplication' in schools and were expected to achieve perfect accuracy irrespective of the situation to which it was applied.

For example, if asked the weight of 285 bricks each weighing on average 0.73 kg, then the precise answer, 208.05 kg, would be expected and 208.04 kg might be marked wrong. Yet, for all practical purposes, the 0.01 difference would scarcely be of importance: 0.01 kg, or 10g, is about the weight of a £1 coin and is hardly significant compared with the overall weight of a load of bricks. In the real world, faced with this calculation it might, for example, be appropriate to approximate the two numbers to 300 bricks (rounding up) and 0.7 kg (rounding down) and to give an approximate answer of 210 kg. However, a builders' merchant might be able to look at the pile of bricks and estimate their total weight as 200 kg, using previous experience to help the estimation.

Notice that in the last paragraph both of the words *approximation* and *estimation* were used. These two words are often confused with each other, but in fact there is a difference which the bricks example might help to illustrate. When approximating, you deliberately lose the accuracy present in the data whereas when you estimate you are making an informed guess, usually trying to be as accurate as possible. When approximating, you move from one number to another (usually) less accurate one: in the example above, there was movement from 285 bricks to 300. When you estimate, you move from a context to a number, just as the builders' merchant might move from the context of the weight of the pile of bricks to the number 200. The distinction might also be clarified by thinking in terms of money. Suppose you estimate the cost of a day trip to France, judging that it will be 'about £200'. The actual cost might turn out to be £223.45. Later when you are asked by a friend how much the trip cost you may approximate and say 'about £220'.

In recent years, the importance of estimation has been stressed in schools much more than previously and it is has frequently been mentioned in various UK curriculum documents. Here are some examples:

make sensible estimates of a range of measures in relation to everyday objects;

use estimation to check calculations;

estimate the number of apples in a bag;

estimate the time it will take to clear away the PE equipment;

estimate $\frac{3}{4}$ of the length of a piece of wood;

estimate the capacity of a teacup;

estimate the 'weight' of a schoolbag.

Notice that most of these examples refer to estimates of *measurements* whereas one, the number of apples in a bag, refers to an estimate of a *count*. This distinction is an important one and you might like to give a few moments thought to it before going any further.

▶ **What do you think is the distinction between measurement and counting?**

All of these estimates involve attaching numbers to certain *variables*. A variable is an important idea in mathematics. It refers to any quantity that can take different values—for example, the number of apples, the time taken to complete a task, the length of a piece of wood, and so on. Later in the course, you will become used to using letters to represent variables: the essence of algebra. In this section, however, you should focus on an important statistical distinction between *discrete variables* and *continuous variables*. Variables which involve counting are known as discrete variables, whereas those which involve measuring are called continuous. With counts (such as the number of apples in a bag), it is possible to be exact, whereas with measurements (such as the weight of a schoolbag), it is never possible to achieve perfect accuracy, as the next activity illustrates.

Activity 15 *Exact measurement?*

A4 is one of the commonest standard sizes of paper (and is the size of the pages in the MU120 assignment booklets, for example).

Measure the width of a piece of A4 paper with a metric ruler and write down your answer.

You should do this before reading any further.

You were deliberately not told in this activity what level of accuracy to use. You might have written down 21 cm and this would carry the implication that the page width was nearer 21 cm than 20 cm or 22 cm.

If you used imperial units, you might have recorded 8.3 inches.

Alternatively, you might have tried to be more accurate and have written down 21.1 cm or 211 mm. Again there would be an implication that the actual measurement was nearer 211 mm than 210 mm or 212 mm.

If you had had access to a more accurate measuring device still, you might have been able to write down 211.0 mm or 211.03 mm or .... . However, no matter how good your measuring device, you would never be able to say what the *exact* width of the particular sheet of paper was. The same is true of all measurements of continuous variables: time, weight, temperature, volume, or whatever.

It also follows that no two sheets of A4 paper can be said to be exactly the same width. However, measurement of their width will inevitably be carried out to a certain degree of accuracy, so that the stated widths of any two sheets might be identical.

Figure 8 should help to give you a mental image of how discrete variables differ from continuous variables. In this case, the discrete scale is shown as a set of unit steps while the continuous scale is represented by a steady slope. An in-between value such as 2.36 is not possible on a discrete scale, but does exist on a continuous one: you cannot have 2.36 guillemots, but you can have 2.36 kg of flour.

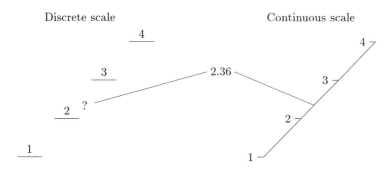

Figure 8   Steps or slope?

In general, then, discrete variables are counted and continuous variables are measured.

Activity 16   *Discrete or continuous?*

Which of the following are examples of continuous variables and which are discrete?

  Price of a loaf of bread in pence.

  Number of guillemots on a cliff face.

  Time of an athlete running 1500 metres, in seconds.

  Goals scored for and against a hockey team.

Distance between major cities, in miles.

TMA score achieved by a student.

Air temperature at midday at a weather station in °C.

Wind speed measured in kilometres per hour.

Wind speed on the Beaufort scale (for example, gale force 8).

The next two subsections, involve carrying out investigations involving estimation; the first involves data arising from a discrete variable and the second data arising from continuous variables.

3.2 Estimating how many

In the video band of the previous section, you saw how guillemots are counted on the island of Skomer, and you heard how an estimate for each cliff face is prepared. Each person who provides counts is in fact using a combination of counting and estimation and, with experience, these counts become more accurate. In this investigation, there will be an opportunity to try this out for yourself, replicating the actual exercise in order to re-create the collection of data in the last subsection. However, the main reason for asking you to do this is that the exercise will raise a variety of statistical issues which will be useful for you to experience in a direct and personal way. In particular, it is an opportunity to emphasize the importance of a well-defined question.

Perhaps when you watched the video you asked yourself how easy *you* would have found it to estimate the number of birds on a cliff face, especially if you were not standing on dry land but were instead in a moving boat. You, like most of the students who follow this course, will probably not have tried counting seabirds before, so how good do you think first estimates would be? It is possible to carry out an investigation to explore this.

◁ *Posing* ▷   Someone might pose the following question for this investigation.

How good am I at estimating numbers of seabirds?

Activity 17   Refining the question

Give some thought to the above question. Is it precise enough for you to be able to move on to the next stage of the investigation and begin to collect some data? What data would you collect? Write down any problems that occur to you and note ways in which the question might be refined in order to provide a practical way of collecting the data.

What ideas from the unit have you used to answer these questions? Add to the response sheet for Activity 6.

The problem with the question as posed at the moment is that it is very general indeed. How can you assess how good you are or someone else is at estimating numbers of seabirds? The following is just one possible approach to this problem.

You could assess someone's ability to estimate numbers of seabirds by making a comparison between their ability and that of other people. Few are likely to be as good at it as the warden on Skomer because of lack of experience, but it may be possible to compare someone's ability with that of other novices. So the question might be refined to the following.

> How do I compare with other people at estimating numbers of seabirds?

One way of making this comparison would be to devise some sort of test to find out how close someone's own estimates are to a known number of seabirds and then compare the test result with those of other people. For example, you might use a particular cliff face such as the Wick as your test site. So you could refine your question further.

> How do I compare with other people at estimating the number of seabirds on the Wick?

You may decide (for various reasons) not to carry out this activity yourself. Try to find someone who is willing to generate an estimate.

It would be lovely to be able to go to the Wick on Skomer to do this, along with lots of other people (perhaps all MU120 students) for comparison. However, there would be one or two problems! One of the problems is that the birds do not stay still and, as the video showed, the number varies from day to day and as the breeding season progresses. Everyone might be visiting the same cliff face, but it is unlikely that they would all be trying to count the same number of birds. The test would not be a standardized one.

One way of providing a more standardized test is by being prepared to move a little away from reality. This could be done by watching a video sequence, showing birds on the Wick. In this way, everyone would look at the same cliff, filmed at the same time, and the test would be a more standardized one.

However, the sacrifice that has to be made is that you would then really be testing people's ability to estimate numbers of birds when seated looking at their television screens, rather than when standing on a windswept cliff in Wales.

(This dilemma is one which has been faced many times before by those concerned with setting tests and examinations. For example, if you wished to find out how good students are at using and doing mathematics, and you wanted to measure their performance, you might decide to use a standardized test or exam. This has the advantage of being fair—all students do the same test—but it is artificial, because it is removed from the business of using and doing mathematics in the real world. It merely provides a measure of how good the students are at doing the exam.)

However, for the purposes of this investigation, using video seems the best that can be done. There is a short video sequence, following the long one you watched for Section 2, lasting about two minutes, which was filmed from the cliff opposite the Wick. This could be used as the test. The camera panned from left to right, following the line of one particular ledge on this huge cliff: viewing this video sequence will allow you to form an estimate of the number of birds shown on the cliff face.

You will also need to decide precisely which birds are to be counted. On Skomer, guillemots and razorbills are counted separately but, because the two look very similar, for this investigation count them all together. At first, the video sequence shows a large number of birds confined to one long ledge on the cliff, but further on there are other guillemots and razorbills sitting above the main ledge, which also need to be counted. The video also shows some kittiwakes which look very different: they are small gulls and their most striking characteristic is their white heads. The kittiwakes and any birds which fly by are not to be included in the count. Figure 9 will help to make this clear.

So once again the question has been refined; it is now as follows.

> How do I compare with other people at estimating the number of guillemots and razorbills sitting on the ledge on the Wick shown in the video sequence?

Activity 18   Can I collect the data?

◁ *Collecting* ▷

Think again carefully about the question driving this investigation. Are there still problems connected with it? What data are needed and are you ready to begin the collection stage now?

Add your comments to the printed sheet used in Activity 14.

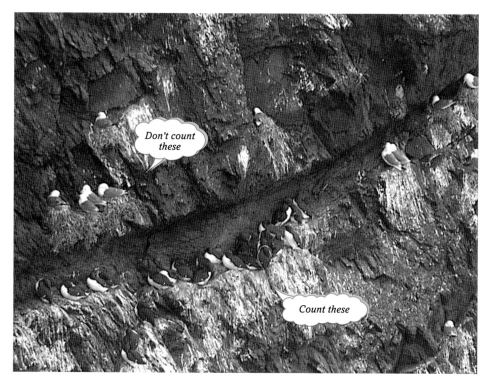

Figure 9   Guillemots, razorbills and kittiwakes

A fair, standardized test is now established, but there are several problems still outstanding.

First, it is all very well having a test, but what is the actual number? If you are to know how good an estimate is, you need to know how many birds there actually were on the cliff, or rather how many are shown in the video sequence. There is no way that anyone can be absolutely certain about the actual number of birds on a cliff face at any time. All you can do is try to get as accurate an estimate as possible. It was felt that the best way to do this was to ask one of the wardens to view the video and to use his value as the best possible estimate. That figure will remain a secret for the moment!

Second, there is the question of the manner in which you watch the short video sequence. Is it all right to use the pause button on the video player or to replay the sequence? Perhaps the test conditions should be made as life-like as possible.

The counters on Skomer would move their binoculars or telescopes, working their way along the ledge in the way that the video camera moved from left to right along the ledge. They might do this two or three times to allow a number of separate estimates to be made, so replicate this in the test: you can view the sequence more than once and produce successive estimates if you so wish. However, pausing the video sequence is not allowed as there is no obvious comparison with the real situation. Approaching the television screen or pointing at particular parts of it are also ruled out.

It is then for you to decide how to use these different figures to produce your own personal estimate. Will you calculate an average (which?) or assume that your later estimates are more accurate than your first ones? You can decide for yourself and commit as little or as much time to the test as you wish.

So you can now carry out the test and collect the data. A number of OU students were asked to carry out the test and their scores, together with the warden's estimate, are provided later in this subsection. These are the secondary data for the investigation. But before you look at them you will need to collect your own primary data: that is, you will need to sit the test yourself and/or invite other people to join you to carry out this activity. If you do, be sure to warn them that it is not going to be easy!

The distinction between primary and secondary data was discussed in Subsection 1.1.

Activity 19   *How many birds?*

Watch the short sequence in band 4b of DVD00107 and make an estimate of the number of seabirds shown sitting on ledges of the cliff face. You may play the sequence more than once if you wish. If you do this, make a note of the separate estimates and decide how to use them to form a final estimate for the number of birds.

You may well be keen to see how 'your' count or estimate compares with that of the warden and with other OU students, but before that, make some brief notes about the process of counting or estimating.

Activity 20   *Count or estimate?*

How did you go about the exercise? Did you count in ones, twos, fives, or what? How did you keep track of which of the birds you had already counted? How many times did you count? How far do you think you were estimating and how far counting precisely? How confident do you feel about your answer? Were you well prepared—for example, did you have a recording sheet handy? Would you use the same method again? If not, how would you modify it? Note your comments for future investigations.

Tricia, one of the counters on Skomer, described how she tackled the counting process, as follows.

> I usually say the numbers to myself in my head and wherever possible I'm counting one at a time. I try to think out a strategy before starting to count—I'll do that little bit of cliff first, then the long ledge, then the odd birds above it, and so on. Where the birds are very closely packed together I sometimes find myself counting in twos or even fives ... 80, 85, 90, 95. All the time, while I'm saying the numbers in my head, I'm also trying to keep a grip on the strategy so that I know which ones I've counted so far.

One problem I often have with a big count is keeping track of the hundreds. I'll be counting the 299th bird and I've just been saying to myself 97, 98, 99, and then it's hard to remember whether it should be three hundred next, or two hundred or four hundred.

One trick I learned from Steve is to fold away successive fingers holding the binoculars to record the number of hundreds. When I get to the end of a big count I usually feel a considerable amount of uncertainty—there's also relief that I've got to the end. It's really hard mental work—and I thought it was easy to count!

Tricia's finger-folding strategy was appropriate because of the circumstances in which she was counting—she was unable to free a hand from holding the binoculars to record in any other way. In other circumstances, different techniques have evolved. Cricket umpires sometimes transfer stones from one pocket to another as each ball of an over is bowled. Knitters have various strategies for keeping count of the number of rows they have knitted: for example, one method is to transfer an object such as a box of matches between the left- and right-hand side of the chair after every five rows.

In the next activity, you move on to an analysis of the primary and secondary data. You will need to work out the error of each person's estimate in relation to the 'right' answer. However, if one estimate is ten too large and another is ten too small, are they equally inaccurate? If you consider this to be so, it would seem sensible to calculate the absolute error for the calculations. The *absolute error* is the amount of the error, paying no attention to whether the error is too large or too small. This would mean that the absolute error for both students would be ten, and this seems sensible. If positive and negative signs are allocated to their errors (+10 and ⁻10), then their average error would be zero and this would obviously be misleading. Absolute errors are treated in more detail in the next subsection.

◁ *Analysing* ▷

Activity 21   Secondary data and analysis

The secondary data for the investigation, the warden's figure and the estimates of twenty other people, are given in Table 6 below.

Table 6   Secondary data

Estimates from twenty students:

| 110 | 134 | 150 | 134 | 113 | 138 | 155 | 125 | 136 | 120 |
| 127 | 130 | 170 | 148 | 124 | 118 | 164 | 135 | 141 | 130 |

Warden's estimate: 188

Analyse the data in order to answer the question posed: work out your own absolute error and that of each of the other people; calculate the average absolute error.

It is noticeable that the OU students tended to underestimate the number of birds, just as inexperienced counters do on the island. It seems that it is much more likely that birds will be missed and not counted at all, than that they will be counted more than once.

▶   Which average should you use?

The choice of median or mean here is really just one of personal preference, as there seem no clear grounds for using one rather than the other. The median is not affected by outliers, but there are no obvious outliers in this batch of data.

◁ *Interpreting* ▷   You should now be able to say whether you are any better or worse than the 'average OU student' at estimating the number of seabirds sitting on the ledge on the Wick shown on the video sequence.

But you may now want to know *how much* better or worse you are than other students. Now you are developing the question a little, so another cycle of the Posing, Collecting, Analysing, and Interpreting stages of a statistical investigation is needed. You would not need to collect any further data, but clearly some work would need to be done at the analysis stage—you would need some way of comparing your error with that of other students.

Activity 22   *How much better (or worse)?*

Suggest a way to compare your error with that of the average OU student. Do the appropriate calculation and interpret the result.

3.3 *Estimating how much*

You are now asked to conduct a two-part investigation into another aspect of your estimation skills. This investigation is based on data arising from continuous variables rather than the discrete variables used in the previous subsection. Both parts of the investigation are described in detail below. You are invited to carry out your investigations as you read about them. It is assumed that you are carrying out this activity while in a room in a building (perhaps a flat or a house). If not, then go on to Section 4 and delay reading about this investigation until you are in a room. You will also need a tape measure for this investigation.

The questions are as follows.
(a) How good are you at estimating heights within a room?
(b) Do your estimation skills improve with practice?

First part of the estimation investigation

How good are you at estimating heights within a room?

◁ *Posing* ▷

Activity 23   Collecting the data

(a) Identify twelve objects in your room and, if the objects are large, choose a particular point on the object (for example, the *bottom* of the light switch, and so on). Your chosen objects should be located at different heights above the floor. Write down the objects in a list.

◁ *Collecting* ▷

(b) Estimate the height (in metres) of each object above the floor.

(c) Using a metric tape measure, measure as accurately as you can the height of each object above the floor and write it down beside the corresponding estimate.

The estimates and measurements produced by Alan, a member of the course team, are given in Table 7.

Table 7   First set of estimates and measured heights

Object	Estimate (m)	Measure (m)
Coffee table top	0.28	0.45
Door handle	0.60	1.07
Light bulb	1.90	1.94
Top of mirror	1.70	1.83
Top of hearth	0.85	0.95
Picture rail (top)	2.05	2.02
Top of door	2.00	1.98
Bottom of light switch	1.15	1.36
Chair seat	0.38	0.51
Mantelpiece	0.60	1.02
Wastepaper basket rim	0.30	0.27
Curtain rail	2.25	2.30

The table contains *paired* data, so a scatterplot would be a useful form of representation.

◁ *Analysing* ▷

Activity 24   Analysing the data

Draw a scatterplot of your own data.

Figure 10 shows Alan's data displayed in a scatterplot. A diagonal line has been drawn on Alan's scatterplot. This line could be called the 'perfect estimation line', since it joins all the points where the estimate would be exactly the same as the measurement. Note that nine of Alan's points lie below the line and three lie above it, indicating that he had a tendency to underestimate heights; at least, heights measured in metres.

◁ *Interpreting* ▷

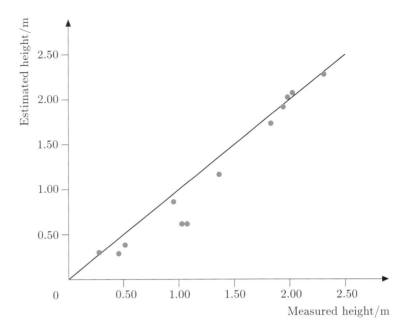

Figure 10   Scatterplot of Alan's first estimates and the measured heights

Alan's most accurate estimates were for heights of around 2 metres, but he felt he was poor at estimating heights between 0.5 m and 1.5 m. He seems to have a good sense of what 2 m is (roughly the height of a door).

Activity 25   *Interpreting*

Draw the 'perfect estimation line' on your scatterplot. Write a few sentences interpreting your own estimation skills from your scatterplot.

Second part of the estimation investigation

◁ *Posing* ▷   Do your estimation skills improve with practice?

◁ *Collecting* ▷   Before repeating the experience, Alan felt he needed to improve his estimating skills, especially for heights in the range 0.5 m to 1.5 m. He thought he might find it helpful to imagine a standard ruler of 30 cm. He also remembered what his own height was in metric units. Make a note of one or two measurements with which you are familiar and which might provide useful benchmarks for your estimations.

Gillian, another member of the course team, when trying out the same activity, measured her stride and the reach of her outspread fingers in metres. She also recalled an informal way of estimating lengths of cloth or string—the distance from one's nose to the outstretched finger tips is about a yard, or a little less than a metre.

Alan repeated the experiment in a different room. The resulting data are shown in Table 8.

Table 8   Second set of estimates and measured heights

Object	Estimate (m)	Measure (m)
Work surface	1.15	0.90
Cupboard handle	1.60	1.60
Top of socket	1.35	1.29
Top of cupboard	2.10	2.07
Chair strut	0.26	0.26
Top of radiator	0.90	0.78
Top of oven hood	1.70	1.73
Radio handle	1.30	1.23
Onion on rack	0.30	0.29
Lamp base	1.45	1.14
Base of spaghetti jar	1.75	1.78
Top of kettle	1.40	1.13

Activity 26   *Practice and more collecting*

Carry out your own short skills session and repeat the experiment in a different room yourself.

Activity 27   *Analysis again*

Draw a scatterplot of your estimates against measured heights for your second set of objects. Comment on whether you have improved your estimating skills.

The scatterplot for Alan's estimated and measured heights is shown in Figure 11.

◁ *Analysing* ▷

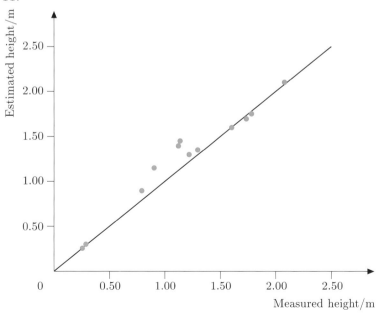

Figure 11   Scatterplot of the data in Table 8

◁ *Interpreting* ▷ Alan's second scatterplot does seem to indicate a slight improvement in his estimation skills. This time eight of the points lie above the perfect estimation line, two below it and two exactly on the line. Alan felt that he had rather overcompensated with his estimates to avoid underestimating. He thought that his estimates within the 0.5 m to 1.5 m were still rather poor, but that his estimates of low heights had improved. Overall, the points seemed to lie closer to the diagonal line than before, suggesting that an improvement had taken place.

Activity 28 *Interpreting again*

Write a few sentences interpreting your own scatterplots.

◁ *Analysing* ▷ Sometimes, at the interpretation stage of an investigation, it is worth asking whether an alternative type of analysis might provide further insights. In this example, calculating the *absolute errors* of the estimates might be a fruitful approach. This idea was introduced in Subsection 3.2, but here it is dealt with more formally.

The absolute values of the estimation errors can be calculated as follows.

$$\text{Absolute error} = |\text{Estimated height} - \text{Measured height}|$$

These two vertical lines are a piece of mathematical notation which means 'take the size of the difference and ignore the sign (+ or –)'. It is read as the modulus.

In other words, the absolute error is the *modulus* of the difference between 'Estimated height' and 'Measured height', and it is the difference ignoring any minus sign in the result. For example, the first error is 0.28 m − 0.45 m, which equals ⁻0.17 m. Ignoring the minus sign, the absolute error is 0.17 m. The complete sets of absolute errors are shown in Table 9.

Table 9 Absolute errors for the first and second set of Alan's estimates

First set of absolute errors (m)	Second set of absolute errors (m)
0.17	0.25
0.47	0.00
0.04	0.06
0.13	0.03
0.10	0.00
0.03	0.12
0.02	0.03
0.21	0.07
0.13	0.01
0.42	0.31
0.03	0.03
0.05	0.27

Alan then keyed these data into his calculator and obtained the summary statistics given in Table 10.

Table 10   Summary statistics for the two sets of errors

	Errors: first estimates (m)	Errors: second estimates (m)
Mean	0.15	0.10
Median	0.12	0.05
Standard deviation	0.14	0.11
Interquartile range	$0.19 - 0.04 = 0.15$	$0.19 - 0.02 = 0.17$

Next, Alan drew boxplots for the two sets of errors. These are shown in Figure 12.

Errors: first estimate

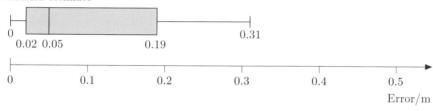

Errors: second estimate

Figure 12   Boxplots of the two sets of absolute errors

Activity 29   *Analysis of absolute errors*

Calculate the absolute errors for your two sets of data. Use your calculator to find summary statistics and obtain boxplots for the absolute errors.

The summary statistics and the boxplots for Alan's data show that not only did the average size of the errors decrease, but the spread of errors was also reduced as measured by the standard deviation or the range. However, note that the interquartile range is greater for the *second* set of estimates. Alan felt that this suggested that he had become a better but no more consistent estimator of heights in metres.

◁ *Interpreting* ▷

Activity 30   *Over to you*

Interpret the summary statistics and boxplots for your two sets of errors.

Outcomes

After studying this section, you should be able to:

◇   understand some of the issues involved in estimation and approximation (Activities 15, 17 to 20);

◇   understand the distinction between data arising from discrete and continuous variables (Activity 16);

◇   understand how the four-stage model is applied to statistical investigations and use it for your own investigations (Activities 17 to 29);

◇   work with 'absolute errors' (Activities 29 and 30).

Activity 31   Defining terms

Before leaving this section, there are a few important terms that you may wish to include as Handbook terms. For example:

estimation, variables (discrete and continuous), approximation, absolute errors, modulus.

Include a short definition and possible use of each term. Check whether there are any other terms you want to include now, and continue to add definitions as you complete the unit.

Think about your use of the hand book. Do you prefer to define terms as you come to them or complete the sheet towards the end of the unit? Which approach is most helpful to you? There is a printed response sheet for this activity.

4 Four case studies

Aims   The main aim of this section is to illustrate, by means of case studies, both the four-stage model of an investigation and the three investigation types that were described in Section 1.   ◇

The section consists of four short case studies which have been written up using the four-stage model, with which you should by now be familiar, namely:

1   Pose the question
2   Collect the data
3   Analyse the data
4   Interpret the results

The case studies have been provided in order to give an indication of the level of detail that you might be expected to produce for your TMA investigation.

These case studies cover the three types of investigation—*Summarizing* (S), *Comparing* (C) and *Looking for relationships* (R), as well as the two sources of data—primary and secondary. These types are listed in Table 11 below. What has not been detailed for each of the case studies is whether the data used are discrete or continuous (see page 37).

Table 11   A set of four case studies

	Type of investigation			Source of data	
	S	C	R	Primary	Secondary
Film review	✓	✓			✓
Famous Five	✓	✓		✓	
Reading ages	✓		✓	✓	
Blanket of clouds	✓	✓			✓

Activity 32   *Case studying!*

As you read through each case study, make a note of whether the data arose from a discrete or a continuous variable. Only then refer to the comments. Read the case studies now, focusing particularly on whether the four-stage model has helped to clarify the main stages of the investigation.

4.1 Film review

Pose the question

◁ *Posing* ▷   One of the members of the course team is a regular *Guardian* reader and is particularly interested in the reviews of films which are about to be shown on television. Over a period of time, he gained the impression that more older films were being selected for review than newer ones. He decided to carry out a statistical investigation to see if this was in fact the case. The question was posed as the following.

Does the *Guardian* TV reviewer tend to select older films?

Collect the data

◁ *Collecting* ▷   He decided to look at the films that were shown on UK television over a particular week. Data on the years in which the films were made were collected from the TV Guide of the *Guardian* of 9 October 1993. These are shown, sorted into ascending order, in Table 12. Note that all the years have been abbreviated so that, for example, 1931 is shown as 31.

Table 12   Years in which the films were made

Reviewed films	31 32 36 36 37 38 40 43 50 55
	56 59 63 69 75 85 85 85 88 90 92
Non-reviewed films	31 32 36 36 37 38 40 41 41 43 43 44 47
	50 52 55 56 58 59 63 69 71 74 75 79 83
	84 85 85 85 85 87 87 87 88 88 89 90 92

Analyse the data

◁ *Analysing* ▷   Summary statistics for the two batches of data were obtained. These are given in Table 13.

Table 13   Summary of the data in Table 12

Statistic	Reviewed films	Non-reviewed films
mean	59.286 (59)	63.718 (64)
standard deviation	21.279	20.696
min	31	31
Q1	37.5	43
median	56	63
Q3	85	85
max	92	92
n	21	39

Most calculators provide two possible values for the standard deviation. The values given here are those with the divisor 'n'.

Comparing means shows that the average year for reviewed films during the week sampled is approximately 1959, whereas the average year for those films not selected by the *Guardian* reviewer is 1964. There is a five-year difference between the means of the data sets, suggesting that the Guardian reviewer does tend to choose older films, but the difference does not seem to be very large. The values for the standard deviations confirm that both sets of data have roughly the same spread. Using the interquartile range as a measure of spread, the reviewed films have a slightly larger spread than that of the non-reviewed films (47.5 years compared with 42 years).

The summary statistics show that both sets of observations have the same earliest and latest year but, although the upper quartiles are the same, the lower quartiles and medians differ.

The mean does not seem to be an ideal calculation here, as it produces answers in decimal years which seem inappropriate in the context (for example, the mean year for reviewed films is 1959.286). The median seems more sensible. The median year for the reviewed films is 1956 and the median year for the other films is 1963. This produces a larger, though still relatively small, difference compared with the difference between the means of each batch of data. The lower quartiles also differ slightly, although again, it is not a very large difference.

A useful way to compare two batches of data of this sort is to draw boxplots for the data. The two boxplots are shown in Figure 13.

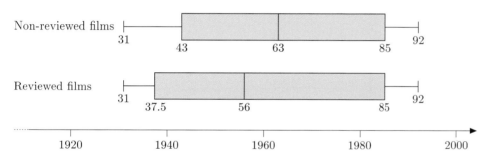

Figure 13   Boxplots of the data from Table 13

The boxplots merely confirm what was apparent from looking at the summary statistics. Although one can tell that the reviewed films are, on average, older than the non-reviewed films, there is only a small difference overall.

Interpret the results

In conclusion, the films chosen by the *Guardian* reviewer do appear to be slightly older than those that have not been reviewed, but this difference does not seem to be very large.

◁ *Interpreting* ▷

4.2   The 'Famous Five' and the mysterious author

Pose the question

As a long-time Enid Blyton fan, I thought I had pretty well the complete collection of her 'Famous Five' series of books. While browsing in a second-hand book shop, I was surprised to unearth a copy of *The Famous Five and the Inca God*. This was not one of her stories that I was familiar

◁ *Posing* ▷

with and the layout and style of presentation looked more modern than the others.

I bought it and, on reading it, decided that, first, it was rather boring and, second, it did not seem to be written in the inimitable Blyton style. I checked inside the front cover but could not find any clear reference to the author. The key question was as follows.

Was this book written by Enid Blyton?

Collect the data

◁ *Collecting* ▷ I decided to use sentence lengths (the number of words per sentence) to investigate authorship. I compared the lengths of fifty sentences chosen at random from this book with the lengths of fifty chosen from a genuine Blyton original from my collection, *Five have a Wonderful Time*. The data are shown in Tables 14 and 15.

Table 14   Lengths of fifty sentences taken from *The Famous Five and the Inca God*

3	5	5	6	6	6	6	7	7	8	8	8	8	9	9	10	11
11	11	12	12	12	12	13	13	14	14	15	15	15	15	15	15	16
16	16	17	18	19	19	20	21	22	23	23	25	27	28	34	35	

Table 15   Lengths of fifty sentences taken from *Five have a Wonderful Time*

4	5	5	5	5	5	6	6	6	6	6	7	7	7	7	8	8
8	8	8	8	8	9	9	9	9	10	10	12	13	13	13	14	15
15	15	16	16	16	17	17	20	21	21	21	22	25	27	28	31	

Analyse the data

◁ *Analysing* ▷ Boxplots were drawn to represent the data as shown in Figure 14.

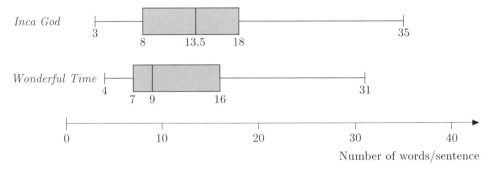

Figure 14   Boxplots showing sentence lengths from the two books

Interpret the results

◁ *Interpreting* ▷ The boxplots recorded small differences in the two batches of sentence lengths. Those from *The Inca God* were more widely spread and contained a number of longer sentences. Enid Blyton's style is known to have been very much based on short sentences. About 50% of the sentences taken from *Five have a Wonderful Time* contained nine words or fewer, compared with an equivalent figure of thirteen words for *The Inca God*.

Just to check that my *Wonderful Time* sample was typical, I counted the numbers of words in fifty sentences chosen at random from another guaranteed Blyton original, *Go ahead, Secret Seven*, which showed the same basic pattern, emphasizing short sentences. A boxplot for these sentence lengths is shown in Figure 15.

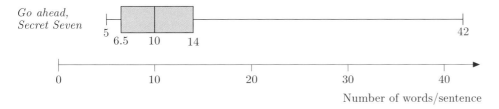

Figure 15   Boxplot showing the lengths of a sample of sentences taken from *Go ahead, Secret Seven*

Although there were some very long sentences in *Go ahead, Secret Seven*, the median and quartile values were all closer to those of *Five have a Wonderful Time* than those of *The Inca God*. This seemed to confirm that genuine Enid Blyton sentences tended to be rather shorter than those contained in *The Inca God*.

A subsequent examination of the small print of this book revealed that *The Inca God* was indeed 'a new adventure created by Enid Blyton, told by Claude Volier, translated by Anthea Bell'. It is unclear exactly what this means; what, precisely, constitutes 'creating' a story? However, this additional information confirms that, as suspected, the actual sentences were not written by Enid Blyton. Whether the writing style of this book was more influenced by the contributions of Claude Volier or Anthea Bell is impossible to say, but it would appear that something got lost in translation!

4.3   Reading ages

Pose the question

A teacher of a class of ten- and eleven-year-olds was interested to see whether the children's reading ages were related to their actual ages. So the central question was as follows.

◁ *Posing* ▷

> Is there any link between pupils' chronological ages and their reading ages?

Collect the data

The children's chronological ages came from the teacher's records and she converted these to 'decimal years'. For example, an age of 10 years and 11 months was entered as 10.9 years, and not as 10.11. She measured their reading ages using the Young cloze reading test and listed reading ages for each member of the class. The data are shown in Table 16.

◁ *Collecting* ▷

Table 16   Ages and reading ages (in years) of pupils in Class 7R

Name	Chronological age	Reading age	Name	Chronological age	Reading age
Andrew	11.2	10.0	Navtej	11.4	8.6
Anthony	11.0	10.5	Nicola	11.2	11.2
Claire B	10.8	11.0	Pardeep	11.6	11.0
Claire E	11.2	9.2	Ravjinder	10.7	9.7
Emma	11.4	13.4	Richard	11.0	11.7
Heather	11.2	9.3	Sandeep	11.4	11.5
Jaspal	11.2	11.7	Sandip	10.9	11.5
Joanna	11.5	11.3	Sarah	10.8	11.1
Joanne	11.6	10.2	Simon	10.7	9.5
Jody	11.5	16.0	Steven	11.0	8.6
Laura	10.9	11.7	Sukhraj	11.0	9.5
Lisa	10.9	12.4	Tanya	11.6	11.8
Mark	11.5	16.0	Tejinder	11.2	9.1
Martin	11.0	16.0	Thomas	10.9	11.1
Meena	11.5	10.0			

Analyse the data

◁ *Analysing* ▷   A scatterplot was drawn of chronological age against reading age, as shown in Figure 16.

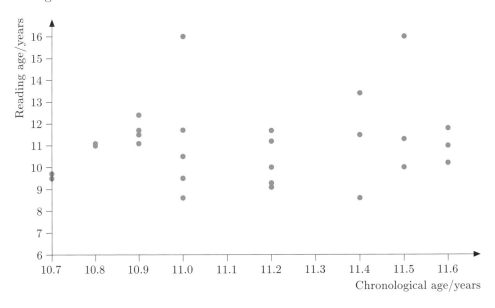

Figure 16   Scatterplot of chronological age against reading age

Interpret the results

◁ *Interpreting* ▷   There is no clear pattern in the scatterplot, suggesting that, for this teacher's class at any rate, there is no link between chronological age and reading age.

4.4 Blanket of clouds

Pose the question

Conventional wisdom suggests that clouds tend to act as a warm blanket, keeping heat in at night and preventing the temperature from dropping too far. I had always accepted this piece of folklore but wondered if it was 'scientific'. The question that I formulated was as follows.

◁ *Posing* ▷

Do clouds help to keep the heat in?

Collect the data

I looked through the various newspapers to see what sort of temperature data were on offer. Data from the *Guardian* seemed ideal. For various towns around the UK, the newspaper gave both the high (H) and the low (L) temperatures for the previous day, and also whether the weather had been sunny, cloudy, raining, and so on. Looking through back copies of the *Guardian*, I was able to find two days of contrasting weather which were reasonably near each other in time. The weather for Monday 28 March 1994 had been largely cloudy and wet, while Monday 2 May was mostly sunny. This meant that I could make a comparison over a number of towns which had experienced cloudy weather on one day, and sunny weather on another day. The data are shown in Tables 17 and 18.

◁ *Collecting* ▷

Table 17   Weather report for 28 March 1994

Around Britain

Report for the 24 hours ended 6pm yesterday

	Sun hrs	Rain in	Temp H	Temp L	Weather (day)
Aberdeen	1.8	.04	11	4	Cloudy
Anglesey	—	.11	10	7	Fog
Aspatria	1.2	.07	11	6	Rain
Aviemore	1.1	.16	10	3	Showers
Belfast	0.2	.13	12	5	Rain pm
Birmingham	0.3	.20	13	5	Rain
Bognor Regis	—	.03	11	7	Drizzle
Bournemouth	—	.22	11	8	Drizzle
Bristol	—	.22	13	8	Drizzle pm
Buxton	—	.21	11	3	Rain
Cardiff	—	.30	11	7	Rain
Clacton	0.3	.03	11	5	Drizzle am
*Cleethorpes					
Colwyn Bay	3.1	.08	15	6	Bright pm
Cromer	—	.07	12	5	Showers
*Dunbar					
Eastbourne	—	.02	11	6	Drizzle
Edinburgh	1.8	.04	13	5	Showers
Eskdalemuir	0.1	.62	7	5	Rain
Exmouth	0.1	.23	13	6	Drizzle
Falmouth	—	.08	12	8	Drizzle
Folkestone	—	.02	10	6	Rain
Glasgow	0.7	.22	11	6	Rain
Guernsey	—	.06	13	9	Cloudy
Hastings	—	.05	10	6	Drizzle
Herne Bay	0.8	—	12	6	Cloudy
Hove	—	.02	10	7	Drizzle
Hunstanton	—	.17	12	5	Rain
Isle of Man	1.1	.04	11	6	Cloudy
Isles of Scilly	0.3	.02	12	9	Drizzle am
Jersey	—	.01	12	8	Rain pm
Kinloss	2.5	.01	12	4	Bright
Leeds	0.3	.04	14	6	Cloudy
Lerwick	5.1	.31	7	4	Rain pm
Leuchars	1.3	.09	12	5	Rain
Littlehampton	—	.03	10	8	Drizzle
Liverpool	1.3	.04	14	5	Bright
London	0.1	.01	12	9	Cloudy
Lowestoft	—	.09	12	6	Showers
Manchester	0.2	.02	14	5	Rain
Margate	0.9	.01	12	6	Cloudy
Minehead	1.0	.20	14	9	Drizzle
Morecambe	—	.09	10	6	Rain
Newcastle	4.1	.01	13	6	Sunny pm
*Newquay					
Norwich	1.1	.17	12	6	Rain
Nottingham	—	.12	12	5	Rain
Penzance	0.3	.07	13	8	Drizzle
Plymouth	0.2	.13	11	9	Drizzle
Poole	—	.25	12	8	Rain
Ross-on-Wye	0.1	.17	13	7	Rain
Ryde	—	.08	11	8	Drizzle
Salcombe	—	.26	12	8	Drizzle
Saunton Sands	0.6	.18	12	5	Cloudy
Scarborough	1.0	.03	13	1	Cloudy
Shanklin	—	.10	11	7	Dull
Skegness	—	.32	12	5	Rain
*Southend					
Southport	1.8	.06	14	4	Bright pm
Southsea	—	.04	11	7	Drizzle
Stornoway	0.7	.23	9	5	Rain
Swanage	—	.26	11	9	Drizzle
Teignmouth	0.1	.16	13	9	Rain
Tenby	—	.44	12	8	Rain
Tiree	1.1	.33	10	6	Rain
Torquay	0.1	.22	13	9	Showers
Tynemouth	3.5	—	13	5	Bright pm
Ventnor	—	.06	10	8	Fog
Weston-s-Mare	—	—	13	8	Rain pm
Weymouth	—	.17	11	8	Cloudy
*Wick					

*Reading not available.

Table 18   Weather report for 2 May 1994

Around Britain

Report for the 24 hours ended 6pm yesterday

	Sun hrs	Rain in	Temp H	Temp L	Weather (day)
Aberdeen	0.2	.12	14	5	Rain
Anglesey	6.9	.03	18	10	Bright
Aspatria	2.9	.06	17	11	Rain am
Aviemore	3.8	—	16	9	Cloudy
Belfast	4.0	.07	15	9	Bright
Birmingham	9.7	—	17	7	Sunny
Bognor Regis	11.7	—	14	6	Sunny
Bournemouth	11.6	—	15	7	Sunny
Bristol	11.7	—	18	8	Sunny
Buxton	7.8	—	15	7	Sunny
Cardiff	10.9	—	16	8	Sunny
Clacton	8.6	—	11	7	Sunny
Cleethorpes	5.8	—	13	5	Bright
Colwyn Bay	5.5	.02	19	10	Sunny pm
Cromer	—	—	11	3	Sunny
Dunbar	0.2	.11	13	8	Rain
Eastbourne	11.6	—	12	6	Sunny
Edinburgh	0.1	.13	15	8	Rain
Eskdalemuir	1.8	.11	14	7	Rain
Exmouth	10.5	—	12	8	Sunny
Falmouth	—	—	15	9	Sunny
Folkestone	11.9	—	13	5	Sunny
Glasgow	0.2	.02	15	10	Rain pm
*Guernsey					
Hastings	11.5	—	13	6	Sunny
Herne Bay	11.3	—	12	5	Sunny
Hove	11.0	—	15	6	Sunny
Hunstanton	10.7	—	15	3	Sunny
Isle of Man	6.7	.05	13	9	Sunny
Isles of Scilly	6.7	—	14	9	Sunny
Jersey	11.4	—	19	9	Sunny
Kinloss	4.4	.03	17	10	Bright
Leeds	4.8	—	17	8	Sunny pm
Lerwick	0.3	.06	9	4	Rain pm
*Leuchars					
Littlehampton	11.2	—	14	5	Sunny
Liverpool	5.0	.07	17	9	Sunny pm
London	11.6	—	17	8	Sunny
Lowestoft	12.3	.—	11	5	Sunny
Manchester	6.0	—	17	9	Sunny pm
Margate	11.8	—	12	6	Sunny
Minehead	11.8	—	19	7	Sunny
Morecambe	4.2	—	18	11	Bright pm
Newcastle	0.5	—	17	8	Cloudy
Newquay	10.6	—	15	10	Sunny
Norwich	13.3	—	12	3	Sunny
Nottingham	9.9	—	17	6	Sunny
Penzance	9.6	—	15	10	Sunny
Plymouth	10.7	—	16	8	Sunny
Poole	11.1	—	15	8	Sunny
Ross-on-Wye	10.1	—	19	7	Sunny
Ryde	10.2	—	12	8	Sunny
Salcombe	11.9	—	13	10	Sunny
Saunton Sands	10.9	—	18	5	Sunny
Scarborough	5.5	—	15	5	Bright
Shanklin	10.2	—	12	8	Sunny
Skegness	7.7	—	11	6	Sunny
*Southend					
Southport	4.7	.01	17	10	Sunny pm
Southsea	10.8	—	13	7	Sunny
Stornoway	4.8	.31	12	8	Bright
Swanage	10.6	—	12	9	Sunny
Teignmouth	8.2	—	12	10	Sunny
Tenby	7.6	—	14	9	Sunny
Tiree	2.4	.16	12	8	Rain
Torquay	10.9	—	14	9	Sunny
Tynemouth	1.4	—	13	7	Cloudy arn
Ventnor	10.3	—	12	8	Sunny
Weston-s-Mare	—	—	17	11	Sunny
Weymouth	11.6	—	13	9	Sunny
*Wick					
*Reading not available.					

Analyse the data

◁ *Analysing* ▷   I was interested in finding temperature swings, so I needed to calculate the difference 'High – Low' for a suitable sample of places. I decided to work with twenty which were cloudy on one of the days and sunny on the other day. The first twenty such places and their corresponding temperature changes from one day to the next are shown in Table 19.

Table 19   Temperature changes on a cloudy day and a sunny day for twenty places

Town	Temperature change (°C)		Town	Temperature change (°C)	
	Sunny (May)	Cloudy (March)		Sunny (May)	Cloudy (March)
Birmingham	10	8	Folkestone	8	4
Bournemouth	8	3	Hastings	7	4
Bristol	10	5	Herne Bay	7	6
Buxton	8	8	Hove	9	3
Cardiff	8	4	Hunstanton	12	7
Clacton	4	6	Isle of Man	4	5
Cromer	8	7	Littlehampton	9	2
Eastbourne	6	5	London	9	3
Exmouth	4	7	Lowestoft	6	6
Falmouth	6	4	Margate	6	6

The data in Table 19 were then analysed in several different ways, as described below.

(a) First, I inspected the data to see how often the temperature change was greater on the sunny day and how often it was greater on the cloudy day. This produced the following results.

No. of places where the temperature change was greater on the sunny day = 14

No. of places where the temperature change was greater on the cloudy day = 3

No. of places where the temperature changes were equal on the cloudy day and the sunny day = 3

Total 20.

(b) Then, using a calculator, I calculated the average (mean) temperature change. The following results were obtained.

The mean temperature change for the sunny day was 7.5 °C.

The mean temperature change for the cloudy day was 5.2 °C.

(c) Finally, I drew boxplots for the data.

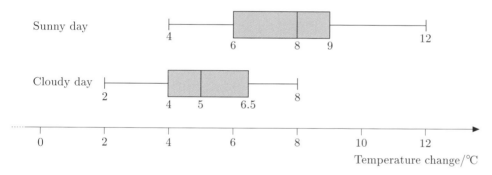

Figure 17   Boxplots showing temperature changes on sunny and cloudy days

Interpret the results

These results do seem to confirm that temperature changes *are* smaller on cloudy than on sunny days. However, there may be several weaknesses with the design of this investigation. First, you are not strictly comparing like with like here, since the sunny days were generally warmer than the cloudy days. So, another possible explanation might be that temperature changes are larger on warmer days. Furthermore, the two batches of data refer to different times of the year. The temperature changes may be more marked when conditions are changing rapidly as summer approaches.

◁ *Interpreting* ▷

Taken together, these design faults might suggest that it would be worthwhile repeating the experiment with more appropriate data.

Activity 33   A marginal activity

Each of the case studies described in this unit has been written up according to the four-stage model and these stages have been flagged in the margin of the text.

Look back over these marginal 'flags' and check that their positioning accords with what you would have expected. Explain to someone else why the 'flags' are there and what they indicate.

Pause briefly to answer the following questions.

◇   How are you keeping to schedule?

◇   Do you need to modify your timetable?

◇   Have you allowed sufficient time for the TMA question?

Outcomes

After studying this section, you should be able to:

◇   classify investigations into the three broad investigation types;

◇   apply the discrete/continuous distinction to variables (Activity 32);

◇   explain how the four-stage model can be used to structure the planning, conduct and write-up of statistical investigations (Activity 33).

5 Tackling your own investigation

Aims   The main aim of this section is to give some general guidance on tackling the investigation in your tutor-marked assignment (TMA).   ◇

One key theme of statistical investigations, stressed in this unit, has been the four-stage model and, in particular, the role of the first stage in establishing a clear question from which the remaining stages will usually follow naturally. In general, it is a good idea for you to gain as much practice as possible at posing questions for yourself and to develop your skill and experience at taking responsibility for all four of the stages.

In particular, when a conclusion is reached, it is important to step back and evaluate the overall design of your investigation. Because this is part of an assignment to be submitted for assessment by your tutor, you will see from its wording that, to some extent, the posing of the question has been done for you. You should still clarify and refine the wording of the question so that it matches as precisely as possible the particular path *you* choose to follow.

Here are some guidelines for helping you to carry out your TMA investigation. Use the four-stage model explicitly in your planning of the investigation and in your write-up.

Pose the question

Clarify and define the central question precisely.   ◁ *Posing* ▷

Collect the data

State what data are required, what sources you have used and describe the   ◁ *Collecting* ▷
nature of the data (for example, whether they arise from discrete or continuous variables). Include your data and any relevant details about how the data were collected, any choices you had to make, and any difficulties you overcame.

Analyse the data

The choice of which statistical techniques (graphical representations or   ◁ *Analysing* ▷
numerical summaries) you decide to use should be made with the following considerations in mind.

(a) Look back to the second stage, collecting the data. The techniques should be appropriate for the type(s) of data involved.

(b) Look ahead to the fourth stage, interpreting the results. The techniques should be suitable for helping you to draw sensible conclusions at the final stage of the investigation.

Interpret the results

The main points to remember here are as follows.   ◁ *Interpreting* ▷

(a) You should attempt to use the techniques from the third stage, analysing the data, to answer the original question posed at the start.

If, having analysed the data, there is no clear pattern, then say so. If you think there is a clear pattern in the data, then be explicit about what criteria you are using for deciding this. Try to come up with a possible explanation for the result.

(b) You should show a healthy scepticism for your conclusions and explore at least one alternative explanation for your findings. You should also take the opportunity to evaluate the overall design of your investigation, consider how it might have been improved, and what you have learned from the experience.

Now that you have nearly reached the end of the unit, check that you have finished Activities 4, 14 and 31 to bring your Learning File material up to date; then complete this final activity.

Activity 34   A final activity

As you have worked through this unit, and indeed the block, you have been asked to think about different aspects of your study. It is not always straightforward to identify strengths and weaknesses in learning and studying, and indeed what you need to do to make improvements. Discussing ideas with other students and your tutor can be helpful and most students find receiving feedback from assignments useful. But often just having some time to think things through and reflect can help to sort out difficulties.

This activity asks you to reflect on the work you have completed so far for Block A and to review how far you have travelled on your learning journey since beginning the course. Think about *what* you have learned and *how* you have learned it. It is quite easy to say 'Yes, I have improved', but what is more difficult is providing evidence that you have improved. For one of the skills you recorded in Activity 4, the skills audit, collect evidence to demonstrate your improvement since starting the course. You may, for example, use responses to some of the activities, or perhaps use an assignment question. Whatever evidence you cite, it should demonstrate your improvement. Finally, think about and plan what you want to achieve during your study of Block B.

Use the printed response sheet if it is helpful.

Outcomes

After studying this section, you should be able to:

◇   use the four-stage model to plan, tackle and write up a statistical investigation of your own;

◇   provide evidence to demonstrate an aspect of improving own learning.

Unit summary and outcomes

In this unit, you have used many data-handling techniques that were introduced earlier in the block, in the context of statistical investigations. A four-stage model for such investigations was presented as a means of providing a structured approach for carrying them out, with particular emphasis being placed upon the need for a well-posed question. You were encouraged to classify investigations into three broad types and you should have found that this helped you to decide which data-handling techniques were most appropriate at the analysis stage of an investigation.

During the unit, you were introduced to several actual investigations, ranging from the large-scale one involving monitoring the size of bird populations on the island of Skomer to the small-scale investigation to establish whether a book had indeed been written by Enid Blyton. Finally, you used the four-stage model to provide a structure within which to tackle your own investigation.

Outcomes

You should now be able to:

◇ explain how a statistical investigation can be broken down into the four stages of posing the question, collecting the data, analysing the data, interpreting the results;

◇ use this four-stage model to plan and carry out an investigation;

◇ take an overview of the statistical techniques used in the analysing stage of an investigation and have some insights into the general purpose of these techniques;

◇ classify investigations into the three broad investigation types;

◇ understand that a large-scale research project such as the monitoring of the size of seabird populations on Skomer consists of a range of statistical investigations;

◇ understand some of the issues involved in estimation and approximation;

◇ understand the distinction between data arising from discrete and continuous variables;

◇ discuss some of the other issues associated with statistical investigations;

◇ provide valid and relevant evidence to demonstrate achievement in particular skills.

Summary of Block A

This block has been concerned principally with statistical ideas. You have been introduced to index numbers (through the RPI and the AEI); summary statistics, such as mean, median and various measures of spread; and graphical representations, such as boxplots and scatterplots. Other concepts discussed included the distinction between discrete and continuous variables and between relative and absolute comparisons. Skills that you should have developed in the course of studying the block include estimating, problem solving and interpreting tables and graphs. You have also learned how to make effective use of material of different kinds—for example, text, audio and video.

One common theme, signalled by the block title 'For better, for worse', has been one of comparison, usually involving aspects of numerical data presented in various means. In the next block, entitled 'Every picture tells a story', a broader range of symbolic representations and mathematical ideas will be introduced.

Comments on Activities

Activity 1

Comments on Activities 1 and 2 use the following numbering: (1) posing; (2) collecting; (3) analysing); (4) interpreting.

(1) Do men earn more than women?

(2) New Earnings Survey data on weekly earnings of men and women (secondary data).

(3) Calculating the earnings ratio at the mean for men and women.

(4) The earnings ratio at the mean shows that women earn less than men on average. However, direct comparisons are difficult to make as men tend to work longer hours than women. You need to take account of the number of hours worked and go round the cycle again with the new data.

For your description, check that the information you have included is accurate and relevant to what is being asked. Is the meaning clear? How have you presented your description–as here in a numbered list—or by using a different format? What helped you decide how you would structure and present your description?

Activity 2

(1) Are boys heavier than girls at birth? Was Shelley's birth weight unusually low? Are the birth weights of girls and boys equally variable?

(2) Birth weights of forty babies were provided by the ward sister (secondary data).

(3) Boxplot values were computed and boxplots drawn to show the range, interquartile range and median birth weights.

(4) The medians indicated that, for this sample, the boys were heavier than girls at birth. Shelley's birth weight was below average but not unusually low.

The spread of boys' birth weights is very slightly greater than that of the girls. An alternative measure of spread, the standard deviation, could be used, which would involve going back to stage 3 of the cycle.

There are many other examples, such as 'A farmer's tale' (*Unit 4*) or investigating gender and occupation (*Unit 3*).

Activity 3

No comments have been included for this activity, because it is important for you to express your understanding of these ideas in your own words. You may like to discuss your entries with your tutor or other students. If you need to lookup terms, use the index at the back of the unit.

Activity 4

An important idea in assessing your own skills achievement is to be able to provide valid and relevant evidence to support your statements. Think carefully what evidence you can present to demonstrate achievement so far, say in communicating mathematics and improving your own learning. Be as specific as you can.

Activity 5

We use the following abbreviations: summarizing (S); comparing (C); looking for relationships (R).

(a) R: this assumes that running speed and length of leg are the two variables. However, it would be C if you were able to look at a long-legged group compared with a short-legged group.

(b) S

(c) C

(d) S

(e) R: for each script, there is an associated pair of scores—one from each examiner.

(f) R: this assumes that income and health can both be measured on some sort of numerical scale.

(g) C: this assumes that death rates are compared before and after compulsory seat-belt laws were applied.

(h) C

Activity 6

In producing your summaries, try to identify the key points and exclude all irrelevant material. Summaries should be concise and clear.

Activity 7

One possible answer would be:
20   40   50   60   90   110   200   400
600   2000

Here the median colony size is $(90+110)/2 = 100$. The total number of birds is 3570.

The larger colonies contain
$110 + 200 + 400 + 600 + 2000 = 3310$ birds.

So 93% of the birds live in larger colonies.

Activity 8

Possible factors affecting the Skomer population include: food availability, breeding success, adult survival rates, oiling and chemical poisoning.

Activity 9

(a) The medians for the last nine of the forty-five sections are given in the table below.

Section	Name	Median
37	Bull Hole	1241.5
38		190
39	Payne's Ledge	219
40	Garland Stone	88
41		20
42		160
43		29
44	Waybench	64.5
45	North Castle	9

The total of the medians is 8696 and this is our estimate of the total number of breeding guillemots on Skomer in 1993.

(b) You may have remarked, for instance, on the fact that two of the largest cliff faces, the Wick and the Amos, only appear to have been counted twice. Was this really so? There are no counts for some sections. Were there no guillemots there? We shall be returning to discuss this further in Subsection 2.5.

(c) The accuracy of the estimate is also discussed in Subsection 2.5.

Activity 10

(a) A plot showing populations of guillemots and razorbills from 1963 to 1993, is given below.

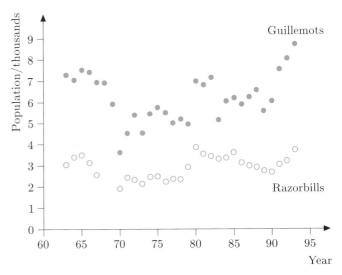

Lines could be drawn between the points in the plot above to help to clarify the trend. However, it would clearly be wrong to try to read off any values from these lines, because the birds are not present for the whole of the year. For example, in any December the population of guillemots on Skomer would be zero.

(b) The guillemot population remained in the region of 7000 during the mid-1960s before falling below 6000 in 1969 and then dramatically to a low point of around 3500 in 1970. Numbers have recovered since then, though not steadily. The population rose to the level of the 1960s by 1991 and by 1993 had reached a figure which is an all-time high.

(c) The great fall was almost certainly associated with the large numbers of birds found dead around the Irish Sea coasts in 1969. This seabird 'wreck' was thought to be due to a severe food shortage. There was a large spillage of oil from a wrecked tanker off Skomer in 1985, but this had only a minor effect on the breeding population. The reason for the recent large increase is not clear, but might indicate a high rate of breeding success together with good food supplies.

(Note in describing the population, precise details are provided wherever possible to support the general statement. In this case dates and population figures are given as evidence.)

(d) In order to make comparisons easy, the razorbill population has been plotted on the same figure as the guillemot population. The graphs are remarkably similar, with many of the peaks and troughs occurring at the same times. Since the two species are so similar in breeding and feeding habits, this strengthens the suppositions that the factors identified did indeed contribute to population changes over the period.

Activity 11

(a) The required statistics are given in the table below.

	Guillemots	Razorbills
Mean	727.8	167.6
Standard deviation (with divisor $n - 1$)	27.6	11.6
Standard deviation (with divisor n)	26.1	11.0
Median	725.5	166.5
Interquartile Range	36	21

Notice that the values of the standard deviations given in Figure 4 are those obtained when the divisor $n - 1$ is used. Notice also that an alternative notation, σ_{n-1} has been employed in the figure. In this course, the standard deviation with a divisor of n rather than $n - 1$ has been used.

(b) See the comments after the activity.

Activity 12

(a) *Guillemots*

$$\text{Relative spread} = \frac{\text{Interquartile range}}{\text{Median}} \times 100\%$$
$$= \frac{36}{725.5} \times 100\%$$
$$\simeq 5\%.$$

Razorbills

$$\text{Relative spread} = \frac{\text{Interquartile range}}{\text{Median}} \times 100\%$$
$$= \frac{21}{166.5} \times 100\%$$
$$\simeq 13\%.$$

(b) See the comments after the activity.

Activity 13

(a) The scatterplot of guillemot numbers against razorbill numbers is given below.

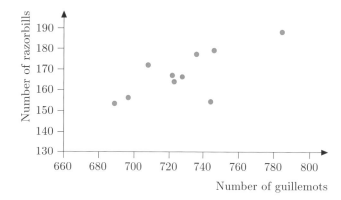

Figure 18

(b) See the comments after the activity.

Activity 14

There are no comments on this activity.

Activity 15

See the comments after the activity.

Activity 16

Discrete:
Prices, numbers of guillemots, goals, TMA score, wind speeds on the Beaufort scale.

Continuous:
Times, distances, air temperatures, wind speeds in km per hour.

Activity 17

See the comments after the activity.

Activity 18

See the comments after the activity.

Activity 19

There are no comments on this activity.

Activity 20

See the comments after the activity.

Activity 21

The students' absolute errors were as follows.

78 54 38 54 75 50 33 63 52 68
61 58 18 40 64 70 24 53 47 58

The mean error is 52.9 and the median error is 54. Notice that it could be in your interests to choose to use the median here, since this suggests a slightly higher average figure for the error in the student estimates.

If your (or your friend's) error was less than these averages you are better than the 'average OU student' at estimating the number of guillemots and razorbills shown on the video sequence sitting on the ledge on the Wick.

See also the comments after the activity.

Activity 22

The mean of the students' errors was 52.9. If, for example, you had an error of 26 you would be justified in claiming to be 'at least twice as good'. On the other hand, an error of 100 would indicate that you were nearly twice as bad as the 'average OU student'. (There is rather more to a comparison of variation than this! But such broad comparisons are sometimes useful.)

Another way of dealing with this might be to look at the spread of errors. A boxplot would indicate that with an error of, say, 40, you could claim to be just better than the lower quartile, whereas with an error of 65 you lie outside the upper quartile.

But in all this, recall that the basis for comparison is results

from only twenty OU students, which is itself a very small sample.

Activities 23–31

There are no further comments on these activities, because they are based on individual data.

Activity 32

Film review	Discrete variable (year of release of films).
Famous Five	Discrete variable (number of words per sentence).
Reading ages	Discrete variable (reading age) and continuous variable (chronological age), although the ages of the children are recorded in discrete one-month steps.
Blanket of clouds	Continuous variable (temperature °C).

Explanation of why reading age is a discrete variable: as might be expected, reading age is based on a count (for example, a count of the number of words correctly read and understood). Although the data might subsequently be manipulated (e.g. using a chart or formula), the resulting reading age can only take certain values within a given range, even if some of those values are decimal numbers. Continuity, on the other hand, requires that all possible values (and therefore an infinite number of values, i.e. measures) are available within a given range. For example, there is an infinite number of chronological ages between, say, 10 years 1 month and 10 years 2 months.

Activity 33

There are no comments on this activity.

Acknowledgements

Grateful acknowledgement is made to the following sources for permission to reproduce material in this unit:

Figure

Figure 4: by permission of Jim Poole, 1993;

Tables

Table 17: the *Guardian*, 29.3.1994, p. 22; Table 18: the *Guardian*, 3.5.1994, p. 22;

Photographs

pp. 17, 22: guillemot and razorbill *Hamlyn Guide to Birds of Britain and Europe*, 1970, Hamlyn;

Cover

Guillemots: RSPB Photo Library; Sellafield newspaper headline: *Independent*, 8.1.1993; other photographs: Mike Levers, Photographic Department, The Open University.

The course team are grateful to the Dyfed Wildlife Trust, Steve Sutcliffe and Jim Poole for providing access to the data used in this unit.

Index